This special symposium volume of the SSHB explores the biological effects of human isolation and migration, and how the situations to which they give rise help to elucidate a variety of biological problems, ranging from evolutionary change to disease etiology. The majority of the case studies presented here are by Asian investigators, and provide an uniquely accessible source of information. This book will be invaluable to those contemplating similar investigations elsewhere, and will be of interest to researchers in a range of disciplines including epidemiology, clinical medicine, demography, anthropology, genetics and evolutionary biology.

SOCIETY FOR THE STUDY OF HUMAN BIOLOGY SYMPOSIUM SERIES: 33

Isolation, migration and health

PUBLISHED SYMPOSIA OF THE SOCIETY FOR THE STUDY OF HUMAN BIOLOGY

1 The Scope of Physical Anthropology and its Place in Academic Studies ED. D. F. ROBERTS AND J. S. WEINER
2 Natural Selection in Human Populations ED. D. F. ROBERTS AND G. A. HARRISON
3 Human Growth ED. J. M. TANNER
4 Genetical Variation in Human Populations ED. G. A. HARRISON
5 Dental Anthropology ED. D. R. BROTHWELL
6 Teaching and Research in Human Biology ED. G. A. HARRISON
7 Human Body Composition, Approaches and Applications ED. D. R. BROTHWELL
8 Skeletal Biology of Earlier Human Populations ED. J. BROZEK
9 Human Ecology in the Tropics ED. J. P. GARLICK AND R. W. J. KEAY
10 Biological Aspects of Demography ED. W. BRASS
11 Human Evolution ED. M. H. DAY
12 Genetic Variation in Britain ED. D. F. ROBERTS AND E. SUNDERLAND
13 Human Variation and Natural Selection ED. D. F. ROBERTS (*Penrose Memorial Volume reprint*)
14 Chromosome Variation in Human Evolution ED. A. J. BOYCE
15 Biology of Human Foetal Growth ED. D. F. ROBERTS
16 Human Ecology in the Tropics ED. J. P. GARLICK AND R. W. J. DAY
17 Physiological Variation and its Genetic Base ED. J. S. WEINER
18 Human Behaviour and Adaptation ED. N. J. BLURTON JONES AND V. REYNOLDS
19 Demographic Patterns in Developed Societies ED. R. W. HIORNS
20 Disease and Urbanisation ED. E. J. CLEGG AND J. P. GARLICK
21 Aspects of Human Evolution ED. C. B. STRINGER
22 Energy and Effort ED. G. A. HARRISON
23 Migration and Mobility ED. A. J. BOYCE
24 Sexual Dimorphism ED. F. NEWCOMBE *et al.*
25 The Biology of Human Ageing ED. A. H. BITTLES AND K. J. COLLINS
26 Capacity for Work in the Tropics ED. K. J. COLLINS AND D. F. ROBERTS
27 Genetic Variation and its Maintenance ED. D. F. ROBERTS AND G. F. DE STEFANO
28 Human Mating Patterns ED. C. G. N. MASCIE-TAYLOR AND A. J. BOYCE
29 The Physiology of Human Growth ED. J. M. TANNER AND M. A. PREECE
30 Diet and Disease ED. G. A. HARRISON AND J. WATERLOW
31 Fertility and Resources ED. J. LANDERS AND V. REYNOLDS
32 Urban Ecology and Health in the Third World ED. L. M. SCHELL, M. T. SMITH AND A. BILSBOROUGH
33 Isolation, Migration and Health ED. D. F. ROBERTS, N. FUJIKI AND K. TORIZUKA
34 Physical Activity and Health ED. N. G. NORGAN

Numbers 1–9 were published by Pergamon Press, Headington Hill Hall, Headington, Oxford OX3 0BY. Numbers 10–24 were published by Taylor & Francis Ltd, 10–14 Macklin Street, London WC2B 5NF. Further details and prices of back-list numbers are available from the Secretary of the Society for the Study of Human Biology.

Isolation, Migration and Health

33rd Symposium Volume of the Society for the Study of Human Biology

EDITED BY

D. F. ROBERTS
Department of Human Genetics
University of Newcastle upon Tyne

N. FUJIKI
Department of Internal Medicine and Medical Genetics
Fukui, Medical School,
Fukui, Japan

AND

K. TORIZUKA
President, Fukui Medical School,
Fukui, Japan

CAMBRIDGE UNIVERSITY PRESS
Cambridge, New York, Melbourne, Madrid, Cape Town, Singapore, São Paulo

Cambridge University Press
The Edinburgh Building, Cambridge CB2 8RU, UK

Published in the United States of America by Cambridge University Press, New York

www.cambridge.org
Information on this title: www.cambridge.org/9780521419123

First published 1992
This digitally printed version 2008

A catalogue record for this publication is available from the British Library

Library of Congress Cataloguing in Publication data
Isolation, migration, and health : special symposium volume of the Society for the
Study of Human Biology / edited by D.F. Roberts, N. Fujiki, and F. Torizuka
p. cm. – (Society for the Study of Human Biology symposium series; 33)
Includes index.
ISBN 0 521 41912 3 (hardback)
1. Medical geography – Congresses. 2. Man – Migrations – Health aspects –
Congresses. 3. Epidemiology – Congresses. I. Roberts, D.F. (Derek Frank)
II. Torizuka, F. III. Fujiki, Norio, 1928– . IV. Series.
[DNLM: 1. Genetics. Population – congresses. 2. Hereditary Diseases – genetics –
congresses. 3. Social Isolation – congresses. 4. Transients and Migrants.
W1 SO861 v. 33 / QZ 50 I85]
RA791.2.I75 1992
614.4′2 – dc20 92-7845 CIP

ISBN 978-0-521-41912-3 hardback
ISBN 978-0-521-06482-8 paperback

Contents

Contributors

T. Arinami Department of Human Genetics, Institute of Basic Medical Sciences, University of Tsukuba, Japan

P.T. Baker 47-450 Lulani Street, Kanoehe, Hawaii, USA

L.A. Bennett Department of Anthropology, Memphis State University, Memphis, Tennessee, USA

B. Bonné-Tamir Department of Human Genetics, Sackler School of Medicine, University of Tel-Aviv, Israel

G. Egusa Second Department of Internal Medicine, Hiroshima University School of Medicine, Japan

N. Fujiki Department of Internal Medicine & Medical Genetics, Fukui Medical School, Matsuokacho, Fukui Pref., Japan

W.Y. Fujimoto Department of Medicine, University of Washington, Seattle, USA

R.M. Garruto National Institutes of Health, Bethesda, Maryland, USA

M. Gushiken Second Department of Internal Medicine, School of Medicine, University of the Ryukyus, Okinawa, Japan

H. Hamaguchi Department of Human Genetics, Institute of Basic Medical Sciences, University of Tsukuba, Japan

H. Hara Second Department of Internal Medicine, Hiroshima University School of Medicine, Hiroshima, Japan

Y. Imaizumi Institute of Population Problems, Ministry of Health and Welfare, Tokyo, Japan

M. Iunes Paulista Medical School, Sao Paulo, Brazil

Y. Kanazawa Omiya Medical Center, Jichi Medical School, Omiya-Shi, Japan

K. Kobayashi Department of Human Genetics, Institute of Basic Medical Sciences, University of Tsukuba, Japan

K. Kondo Department of Public Health, Hokkaido University School of Medicine, Kitaku Sapporo, Japan

P. Lefevre-Witier Centre d'Hemotypologie du CNRS, C.H.U. de Purpan, Avenue de Grande Bretagne, Toulouse, France

K.C. Malhotra, Indian Statistical Institute, 203 Barackpore Trunk Road, Calcutta, India

G. Mimura Second Department of Internal Medicine, School of Medicine, University of the Ryukyus, Nishihara, Okinawa, Japan

R. Miyazaki Department of Medicine, Kudanzaka Hospital, Tokyo, Japan

D.P. Mukherjee Department of Anthropology, University of Calcutta, India

K. Murakami Second Department of Internal Medicine, School of Medicine, University of the Ryukyus, Nishihara, Okinawa, Japan

J.V. Neel Department of Human Genetics, University of Michigan, Ann Arbor, Michigan, USA

S. Nevo Institute of Evolution, University of Haifa, Haifa, Israel

S. Ogawa University of Hawaii, Honolulu, Hawaii, USA

K. Omoto Laboratory of Evolutionary Genetics, National Institute of Genetics, Mishima, Japan

A. Oppenheim Department of Hematology, Hadassah University Hospital, Jerusalem, Israel

S.S. Papiha Department of Human Genetics, University of Newcastle upon Tyne, Newcastle upon Tyne, England

I. Prior University of Otago, Department of Community Health, Wellington School of Medicine, New Zealand

D.F. Roberts Department of Human Genetics, University of Newcastle upon Tyne, Newcastle upon Tyne, England

P. Rudan Institute for Medical Research & Occupational Health, Mose Pijade 158, Zagreb, Yugoslavia

N. Saitou Laboratory of Evolutionary Genetics, National Institute of Genetics, Mishima, Japan

D. Simic Institute for Medical Research & Occupational Health, Mose Pijade 158, Zagreb, Yugoslavia

A. Sujoldzic Institute for Medical Research & Occupational Health, Mose Pijade 158, Zagreb, Yugoslavia

R.I. Sukernik Institute of Cytology and Genetics, Siberian Branch of the USSR Academy of Sciences, Novosibirsk, USSR

A. Ticher Department of Human Genetics, Sackler School of Medicine, University of Tel-Aviv, Israel

K. Tokunaga Blood Transfusion Service, University Hospital, The University of Tokyo, Japan

S. Tsuchiya Institute of Community Health, University of Tsukuba, Japan

F. Vogel Institute für Humangenetik und Anthropologie der Universitat, D-6900 Heidelberg 1, Germany

Y. Watanabe Department of Human Genetics, Institute of Basic Medical Sciences, University of Tsukuba, Japan

M. Yamakido Second Department of Internal Medicine, Hiroshima University School of Medicine, Japan

K. Yamane Second Department of Internal Medicine, Hiroshima University School of Medicine, Japan

Y. Yamanouchi Department of Human Genetics, Institute of Basic Medical Sciences, University of Tsukuba, Japan

H. Yanagi Department of Human Genetics, Institute of Basic Medical Sciences, University of Tsukuba, Japan

R. Yanagihara National Institutes of Health, Bethesda, Maryland, USA

T. Yanase No. 219, Health Care Sawara, Fukuoka City, Japan

N. Yasuda Division of Genetics, National Institute of Radiological Sciences, Chiba-shi, Japan

A. Zoossmann-Diskin Department of Human Genetics, Sackler School of Medicine, University of Tel-Aviv, Israel

Preface

The symposium, held at Phoenix Plaza, Fukui, Japan, July 30-31 1990, to celebrate the 20th anniversary of the International Association of Human Biologists was highly successful. The theme was isolation and migration, for the reasons given in his presidential address by Professor Fujiki. Participants were invited to present their work on the dynamics of these two processes, and their implications for biology and health. The contributions are reproduced in this volume.

The symposium owed its success to many factors. It was the first IAHB conference to be held in the Orient. It was itself part of a week-long joint conference with the Japanese Society of Human Genetics (celebrating its 35th anniversary), preceded by a satellite discussion organised by the Council for International Organisations of Medical Sciences. The proceedings of other discussions during the week on 'Genetics, ethics and human values' and on 'Education and bioethics in medical genetics' have been published under the title 'Medical genetics and society' (1990) edited by N. Fujiki, V. Bulyzhenkov and Z. Bankowski. But much of its success stemmed from the especially cordial relations between the organising committees and the Fukui Medical School, whose members contributed with immense energy and enthusiasm to the week's events. We are glad to record publicly our indebtedness to Fukui Medical School, to all members of the committees responsible for organising the meetings, and particularly to Professor Emeritus E. Matsunaga, former Director of the National Institute of Genetics. Grateful acknowledgement is made for the sponsorship of WHO, UNESCO and CIOMS; the Ministry of Education Japan, the Japan Society for the Promotion of Science; the Fukui Prefectural Government, Fukui International Association, Fukui Municipal Government, Fukui Convention Bureau, Fukui Prefectural Medical Association, and Fukui Municipal Medical Association, the Commemorative Association for the Japan World Exposition (1970); the Uehera, Kashima, and Naito Research Foundations; the Japan China Medical Association; the French Embassy in Japan; the U.S. Man and the Biosphere programme; and the International Union of Biological Sciences.

D. F. Roberts
N. Fujiki
K. Torizuka

1 *The legacy of the IBP: Presidential Address*

NORIO FUJIKI

At the inception of the International Biological Programme in the early 1960s, it was recognised that man's adaptability had led to rapid and far-reaching changes in the environment. The coordinated study of biological variability and management of natural resources promoted through the International Biological Programme was designed to examine the functional relationship of living things to their environment. There was concern both for the needs of man and for greater ecological understanding. The general aims of the human adaptability section of the programme (Collins & Weiner, 1977) were to survey human adaptability in a wide variety of climates, terrains and social groupings, and so to increase understanding of the biological basis of adaptation, and to apply this knowledge to health and welfare problems. In order to implement these goals, an integrated approach was required, utilising standard methods drawn from the many disciplines of human biology.

During the operational phase of the programme, 1967-72, some themes were explored worldwide, namely growth and development, physique and physical fitness, climatic tolerance, genetic constitution, nutrition, and disease susceptibility. Others were on a regional basis. There were teams at work in the circumpolar and cold climates, studying the Eskimo, Lapp, Ainu, and Aleut; in the hot climates, as in central Africa, among Babinga Pygmies, in the African savana, among west African Nigerians, southern and other Bantu peoples, and in east Africa; in high altitudes, as in Chile, the Himalayas, Ethiopia, the Andes and Caucasus. Comparisons between rural and urban communities were the subjects of long-term study in several countries, as in the Kumiyama and Nihonsan study in Japan, the Framingham study in the USA, as well as in Bulgaria, Poland and Czechoslovakia.

The small communities living in a high degree of isolation on islands of Polynesia, Melanesia, Tristan da Cunha, and Japan, or inland in Yugoslavia, in remote marshes on the Danube, and in white Russia and Siberia, were incorporated in IBP studies. There were also investigations on migration from rural to industrialised conditions, for example European immigrants to north America, Yemenite settlers in Israel, Polynesian islanders to New Zealand, Bantu into industrialised cities of South Africa, as well as Japanese Nisei in California, Hawaii and the Pacific northwest.

Today, many of those who participated in these studies have been succeeded by their students. There has been continuing research on many of the IBP topics. In recent years there have been the two programmes on tropical communities, their genetics and working capacity, as part of the IUBS Decade of the Tropics Programme. Another sequel, arising out of the success and interest of the human adaptability section of the IBP, was the foundation of the International Association of Human Biologists. For a symposium to mark its recent 20th anniversary, it was appropriate to choose a theme looking back at the IBP, and looking forward from current to future research. It was important that the theme be somewhat apart from, yet relevant to, the IAHB research activities in the IUBS Decade of the Tropics Programme. The obvious choice was isolation and migration, important topics in the IBP and today of concern in human health programmes. For this theme illustrates clearly the interaction between genetic constitution and environment, as is well shown in some of our own studies.

Isolate studies

Inbreeding effects in isolates

Surveys were carried out in various locations in Japan (Figure 1.1) and these showed consanguinity rates ranging from 8.6 to 58.0%, and about half of consanguineous unions were between first cousins. The mean inbreeding coefficients of sibships for the total populations were high, ranging from 0.00459 to 0.02427 (Table 1.1). Besides these very high levels of mean inbreeding in some locations, those individuals who were not inbred were closely related to other members of the village (Figure 1.2).

The rates in the 20 year period after World War II were compared with those in the twenty year period preceding it (Table 1.2). Though some localities showed a decline in consanguinity rate and mean inbreeding coefficient, others showed an increase. So that the breakdown of geographical isolation is not necessarily paralleled by decline of inbreeding.

There was no clear difference between mean sibship size in the offspring of consanguineous and non-consanguineous unions,

Figure 1.1. Communities surveyed.

although in some of the areas there occurred the phenomenon of reproductive compensation similar to that in Hirado (Schull *et al.*, 1970). This phenomenon, in which the higher infant death rate is compensated by larger sibship size, occurred more often among consanguineous couples, and so makes up for any homozygous disadvantage in their offspring. The infant death rate (mortality during the first year of life) represented a lethal equivalent of 0.26 per gamete, a lower value than in other Japanese data. The extent

Table 1.1. *Breeding structure in isolated communities*

Area	Locality	Pop.	Couples	Inbred couples	≥ 1C	≥ 1.5C	≥ 2C	2C >	Inbred rate	1C rate	Mean α
Islands											
Mishima	H	1975	517	92	52	16	15	9	17.8	10.1	0.00606
	U	737	192	50	29	12	7	2	26.0	15.1	0.01164
Nuwajima	M	1191	340	52	29	8	11	4	15.3	8.5	0.00536
	K	648	175	15	7	2	2	4	8.6	4.0	0.00480
Okishima		686	286	82	40	11	19	12	28.7	14.0	0.01065
Villages											
Midono		161	54	14	8	1	4	1	25.9	14.8	0.02030
Kurodani	Ku	243	83	8	5	3	0	0	9.6	6.0	0.00728
	Ok	136	48	16	8	6	2	0	33.3	16.7	0.01462
Arihara		181	69	40	13	9	11	7	58.0	18.8	0.02427
Mukugawa		235	83	29	6	5	12	6	34.9	7.2	0.00519
Aso		251	95	20	13	5	2	0	21.1	13.7	0.00915
Tomiyama		243	97	38	15	8	10	5	39.2	15.5	0.01556
Miyama	A	269	121	20	11	6	2	1	16.5	9.1	0.00459
	K	607	307	54	20	18	7	9	17.6	6.5	0.00608

Table 1.2. *Chronological change in inbreeding*

Area	Locality	Inbred rate		Mean inbreeding coefficient	
		1925-44	1945-64	1925-44	1945-64
Islands					
Mishima	H	13.7	18.2	0.0085	0.0076
	U	23.2	24.0	0.0179	0.0136
Nuwajima	M	18.0	15.9	0.0022	0.0032
	K	8.5	7.0	0.0032	0.0014
Okishima		35.7	28.0	0.0085	0.0125
Villages					
Midono		27.5	28.5	0.0192	0.0056
Kurodani	Ku	13.2	12.0	0.0073	0.0039
	Ok	30.0	32.0	0.0158	0.0168
Arihara		60.0	58.0	0.0234	0.0313
Mukugawa		34.4	38.5	0.0098	0.0076
Aso		22.0	20.8	0.0193	0.0217
Tomiyama		24.4	52.6	0.0098	0.0208
Miyama	A	19.2	19.9	0.0144	0.0066
	K	20.0	21.3	0.0151	0.0064

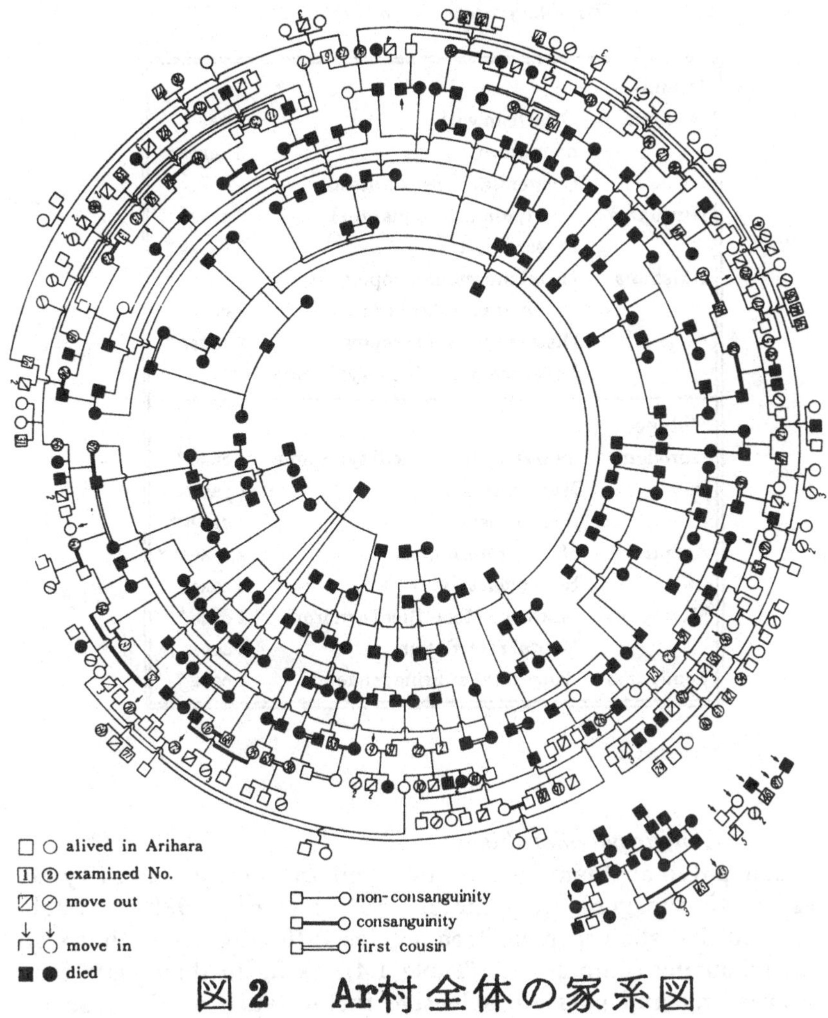

Figure 1.2. Pedigree in Arihara village.

of inbreeding depression as observed in anthropometric data ranged from 0.5 to 2.0%, the higher values tending to occur in the less inbred populations.

Clusters of hereditary diseases or congenital malformations occurred in the highly inbred communities. The particular types varied from village to village, only the Laurence-Moon-Biedl syndrome occurring in several (Table 1.3).

Table 1.3. *Hereditary diseases found in the isolates*

Islands		
Mishima	Dwarfism with	1 case
	neurofibromatosis	3 cases *
	Congenital hyperkeratosis	
Nuwajima	(palmar and plantar)	3 cases
	Pigeon chest	3 cases *
Okishima	Distal muscular atrophy with	
	mental retardation	3 cases
	Distal muscular atrophy	2 cases *
	Laurence-Moon-Biedl syndrome	
Villages		
Kurodani	Laurence-Moon-Biedl syndrome	1 case *
	Marie's ataxia	1 case
	Deafmutism	2 cases *
Arihara	Mental retardation	2 cases* *
	Brachydactyly	2 cases
Mukugawa	Laurence-Moon-Biedl syndrome	1 case *
	Mental retardation	2 cases *
Tomiyama	Congenital afibrinogenaemia	1 case *

* Inbred family

Genetic polymorphisms

Blood polymorphisms were surveyed in each community to establish the gene frequencies (Fujiki *et al.*, 1982). Each community showed gene frequencies differing from those of neighbouring communities (Table 1.4), owing to their isolation, inbreeding and possible selective factors such as the disease pattern. This analysis made it very clear how much biological and social factors influence the characteristics of communities. Differences in geographical distribution, adaptation to environmental factors, genetic drift and founder effect, were among the biological and social consequences of isolation and migration.

Bioassays were attempted making use of phenotype and mating data in a small Japanese population; the results using the koseki records, the detailed family registration system that exists in Japan, could be compared with those from the study of polymorphisms. Inbreeding coefficients can be obtained using the koseki records, as has been done in many studies, but this type of

Table 1.4. *Gene frequencies in the isolates*

Area		*ABO*			*Hp*		*Tf*		*Gc*		*AcP*		*EsD*	
System / *Allele*		*A*	*B*	*O*	*Hp1*	*Hp2*	*TfC*	*TfD*	*Gc1*	*Gc2*	*Pa*	*Pb*	*EsD1*	*EsD2*
Islands														
Mishima		0.267	0.190	0.544	-	-	-	-	-	-	-	-	-	-
Nuwajima		0.247	0.228	0.525	-	-	-	-	-	-	-	-	-	-
Okishima		0.333	0.170	0.497	0.383	0.617	0.998	0.002	0.619	0.381	0.181	0.819	-	-
Villages														
Midono		0.248	0.511	0.241	-	-	-	-	-	-	-	-	-	-
Kurodani	Ku	0.256	0.063	0.680	0.162	0.838	1.000	0.000	-	-	-	-	-	-
	Ok	0.181	0.181	0.639	0.205	0.795	1.000	0.000	-	-	0.079	0.921	-	-
Arihara		0.317	0.198	0.485	0.117	0.883	0.994	0.006	-	-	0.219	0.781	-	-
Mukugawa		0.365	0.137	0.498	0.212	0.788	1.000	0.000	-	-	0.158	0.842	-	-
Aso		0.337	0.154	0.509	0.298	0.702	1.000	0.000	-	-	0.171	0.829	-	-
Tomiyama		0.224	0.227	0.549	0.170	0.830	0.907	0.093	-	-	0.235	0.765	0.638	0.362
Miyama	A	0.398	0.115	0.487	0.220	0.780	0.990	0.010	0.735	0.265	0.294	0.706	0.608	0.392
	K	0.249	0.206	0.545	0.256	0.744	1.000	0.000	0.738	0.262	0.150	0.850	0.687	0.300
Kinki area		0.278	0.182	0.540	0.279	0.721	0.988	0.012	0.737	0.263	0.210	0.790	0.649	0.351

pedigree study does not allow the ascertainment of remote consanguinity. Phenotype bioassay and the correlation method (Yamamoto *et al.*, 1974) provided alternative estimates. In the former the inbreeding coefficient can be estimated from the system deviations from Hardy-Weinberg equilibrium, and in the latter the coefficient of kinship estimated by distance or isonymy. For village Arihara the mean inbreeding coefficient was estimated at 0.0243 by pedigree analysis, 0.0514 by phenotype bioassay and 0.0590 by the correlation method (Table 1.5). The kinship coefficient of random pairs was estimated to contribute nearly half the observed kinship. By way of explanation of this difference, the low estimate from the pedigree study may be attributed to incomplete ascertainment of inbreeding in the ancestors, or the

Table 1.5. *Inbreeding coefficients, village Ar*

a. *From pedigrees*

Generation	*Number*	*Mean inbreeding coefficient*
I	6	0.0104±0.0147
II	47	0.0175±0.0263
III	75	0.0263±0.0342
IV	52	0.0292±0.0319
Total	180	0.0243±0.0316

b. *From phenotype bioassay*

Genetic system	*Inbreeding coefficient*	*Score Uα*	*Information Kα*
ABO	-0.0000±0.1635	-0.000013	37.400
MN	0.1692±0.1105	0.000007	81.962
Haptoglobin	0.2992±0.1463	0.000036	46.693
Acid phosphatase	0.0478±0.1101	-0.000002	82.435
Phosphoglucomutase	-0.0789±0.0838	0.000030	142.496
Combined	0.0949±0.0550		331.090
Excluding Hp & ABO	0.0514±0.0639		245.229

c. *From correlation*

System	*Allele*	*Inbreeding coefficient*	*Gene frequency in the whole Japan**
ABO	A	0.00798	0.2773
	B	0.00511	0.1706
	O	0.01810	0.5521
MN	M	0.28409	0.5433
Haptoglobin	Hp^1	0.12521	0.2750
Acid phosphatase	P^a	0.00009	0.2150
Phosphoglucomutase	$PGM^1{}_1$	0.03844	0.7740
Total		0.06843	
Excluding Hp		0.05897	

* Sources of the data employed:
ABO and MN: Nei and Imaizumi (1966)
Haptoglobin: Matsumoto (1966)
Acid phosphatase: Omoto and Harada (1969)
Phosphoglucomutase: Ishimoto *et al.* (1969)

higher estimates by the bioassay and correlation methods to sampling errors and unknown technical difficulties. The true estimate probably lay between these extreme values. But certainly some factors other than small population size are operating to maintain the high frequency of consanguineous unions.

Besides our studies, and those of Yanase in Shikajima, Shiiba and Okutama, continuous surveys covering the past 30 years have been made in Kishimoto (Inadani), Tsuji (Okuminomote), Handa (Soyadani), Terawaki (Koshikijima) and Shiomi (Goto Islands). Furthermore a large scale survey was made on school children in Shizuoka by the consanguinity study subcommittee of the Science Council of Japan, under the chairmanship of Komai and Tanaka (1973), with the cooperation of some 65 geneticists and physicians. Also there was carried out the child health survey in Hiroshima, Nagasaki and Hirado, directed by Neel and Schull with Japanese collaborators (Komai & Tanaka, 1972; Neel & Schull, 1965), and this involved a follow-up in the years 1960-62 and 1968-72, of the children seen in the consanguinity study carried out under the Atomic Bomb Casualty Commission auspices .

Sociocultural background of consanguinity

The mating patterns in human populations reflect the social structure, and directly influence the gene pool (Mascie-Taylor & Boyce, 1988). The practice of consanguineous marriages in Japan, India, Arabia and other countries is part of each country's history, geography and tradition, and is mainly influenced by sociocultural factors. Japan as an entity is geographically isolated from the neighbouring Asian countries by the natural barriers of the oceans surrounding it. Then the steep mountains create small isolated and inbred compartments throughout Japan, each with its own characteristic culture and unique gene pool. For consanguinity studies the Japanese possess many advantages, e.g. the high rate of consanguineous unions, the relatively complete family registration system of the koseki records, the high fertility, and the low rates of migration associated with the high endogamy rates. However today, with the rapid breakdown of geographic isolation in Japan as in all countries, because of modern developments in transportation and communication, the opportunities are being reduced for research on the homozygotes for many deleterious inherited diseases.

In some isolated communities, consanguineous marriages occur at chance frequencies, depending on the number of mates available (Nakanaga, 1990). This has been seen amongst the Hopi Indians, the Amish, and the islanders of Tristan de Cunha. The tradition of family names in Japan as in much of Europe and in many tribal and caste groups in India means that surname analysis can provide useful information on inbreeding.

The existence of Parsi, Muslim, Christian, Buddist and other religious sects is an important factor making for social isolation and resulting in increased inbreeding. This has, for example, occurred in Kuroshima in Japan, where the Buddists are isolated from the Catholic or Kakure (hidden Christianity) even though the island is so small (Schull *et al.*, 1962). Sometimes lesser but nonetheless important considerations enter into the selection of a mate, for example avoidance of marriage with individuals of a family with a history of certain diseases (social or genetic) or with individuals of a certain parish. Villagers of Kurodani village do not want to marry outsiders because they want to keep in their own hands the secret of their own techniques for paper-making. The long-persisting practice of consanguinity in southern India is mainly due to sociocultural rather than geographical or historical factors. Vedic subunit endogamy restricts the choice of mates to persons within the group and preferably to cousins. In religious and occupational castes of rural and semi-urban areas, the preference for consanguineous unions may be to maintain caste exclusiveness of occupation, to reduce economic instability, to avoid the payment of dowry, to maintain kinship bonds, or to keep inherited property within the family circle. Group solidarity in political and social activities, the perpetuation of the joint family system, and the prohibition of village exogamy, are also related to inbreeding. Avoidance of conflicts between wife and mother-in-law and maintenance of family peace are important in Japanese communities, where the male sex is favoured. Inbreeding can result from demographic patterns of unequal age/sex distribution, while other factors include geographic proximity, economic stability, a desire to expand the circle of relatives, and choice of prospective mate whose relatives know each other well.

The most common explanation, geographical reasons, does not necessarily imply geographical proximity. It has been repeatedly shown that the radius of marriage circle is smaller for consanguineous marriages. Because of the precedence established by the parents, the children of consanguineous couples marry relatives twice as frequently as do the children of non-consanguineous couples. This tendency may lead to higher inbreeding levels in the offspring of consanguineous unions than would be established from the parental relationship alone. While it is generally true that isolation is disappearing rapidly on account of the development of urbanisation and transportation, our data suggest that in some instances in Japan this may not be so.

Besides geographical, religious, economic, and social factors, others which have led in the past to the perpetuation of isolation include political factors, as in the emergence of the Heike villages. These are villages of fugitive warriors who lived in secret, remote and isolated communities in order to escape from a government hostile to them. These villagers, such as those in village Ar, are proud of their royal blood, and do not want to marry outsiders or neighbouring villagers.

The study of disease

The study of genetic polymorphisms (Watanabe *et al.*, 1975; Mourant *et al.*, 1976) has contributed greatly to our knowledge not only of the genetic characteristics of given individuals or populations, their relationships and affinities, but also to our understanding of evolutionary changes of protein molecules and population structures.

Disease susceptibility associated with a given polymorphic

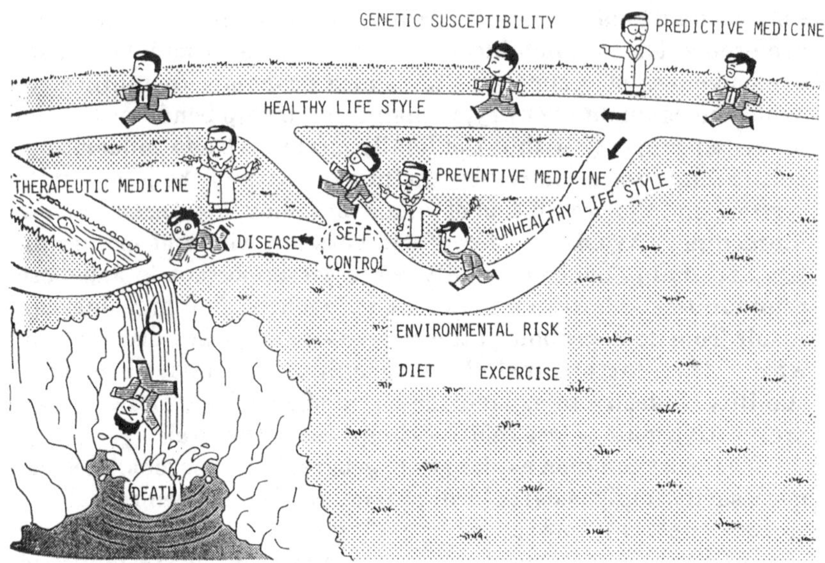

Figure 1.3. Concept of predictive medicine.

trait (Mourant *et al.*, 1978) has helped clarify the etiology of some polygenic diseases. For example a significant association of blood group A and gastric cancer was the first to be established, which suggested the desirability of searching for similar relationships of other blood groups, red cell enzyme and serum protein types. Our preliminary studies suggested a higher than expected risk for leukaemia and aplastic anaemia among persons of haptoglobin type 1-1, and revealed higher frequencies of genes Gc2 and Hp1 in some cancer populations, suggesting genetic susceptibility to malignancy.

These and similar data suggest that there are genetic predispositions as well as environmental agents in etiologies of specific diseases. Multifactorial diseases may possibly be prevented if those who are susceptible avoid exposure to the environmental agents involved. Such multifactorial diseases are assumed to have some threshold delimiting phenotype abnormality, so that those individuals whose genotypic combinations place them near to that threshold are more susceptible to the development of the disorder. Following the discovery of appropriate genetic markers (e.g. the HLA complex and apo-lipoprotein E), an increasing number of persons can be identified who, from their genetically determined constitution, are more susceptible to certain diseases for given environmental

Table 1.6. *Frequency of consanguinity, familial occurrence, and malformations in aplastic anaemia relative to that in controls*

	Aplastic anaemia/ controls	*Fanconi anaemia/ controls*
Consanguinity		
First cousin	1.9	4.2
Closer than the third cousin	1.7	5.0
Familial occurrence		
Aplastic anaemia	3.8	12.5
Severe anaemia	4.0	4.8
Malformations	12.5	62.5
Malignant neoplasms	4.2	13.9
Malformations		
Minor abnormalities	11.2	37.0
Multiple abnormalities	25.0	500.5
Dubois' sign	3.0	8.1

factors such as diet and habit. Here there is the concept of predictive medicine rather than therapeutic or preventive medicine (Figure 1.3). A variety of markers has been studied in coronary heart disease, under our WHO international cooperation programme (Fujiki, 1981).

Knowledge of the biology of disease has been increased not only in the study of genetic polymorphisms, but also by data from migrants. Clinicogenetic studies of aplastic anaemia showed a high frequency of consanguinity (Table 1.6) and a heritability of 47.6%, as compared with that of 37.5% in leukaemia, and also confirmed the close association of polygenic susceptibility, as indicated by genetic polymorphisms, and of high risk environmental factors of diet and habit in the migrant population of Japanese Nisei in Hawaii. There was a mortality rate of aplastic anaemia higher than 0.9 per hundred thousand in Japan, as compared with a European rate of 0.4 per hundred thousand. However the decrease in mortality rate of aplastic anaemia in Japan in the younger generation and the concomitant increase in the older, suggested that changes of lifestyle affected the disease. This suggestion was pursued and supported by a study in the Japanese migrants to Hawaii. The age-adjusted mortality rate

Table 1.7. *Age-adjusted death rates from aplastic anaemia among Japanese in Hawaii and California compared with Japan*

	per 100,000		*Sex ratio*
	Male	*Female*	
Japan (1968-1975)	0.95	1.03	0.89
Hawaii (1968-1975)	0.32	0.31	1.03
California (1968)	0.48	0.50	0.96

amongst Japanese Americans in Hawaii, about 0.3 per hundred thousand in 1968-75 (Table 1.7), was half that for the indigenous Japanese in Japan, and the same as for Europeans. Interestingly, most Japanese in Hawaii had migrated from the south-west part of Japan, where the standardised mortality rate of aplastic anaemia was higher than the national average.

Conclusion

Genetico-anthropo-epidemiological studies help understanding of the causes and natural history of disease. In earlier decades of this century the major communicable diseases were eliminated or controlled. People now live longer and their behaviour and environment have changed in the direction of overnutrition, less physical activity, and greater stress. At the same time, the incidence and prevalence of non-communicable diseases, such as coronary heart disease, hypertension and diabetes have increased. Such trends are to be seen also in many developing countries. The rapid modernisation experienced by such populations has also resulted in unlimited opportunity for the study of the genetic and environmental factors, and the underlying biochemical, behavioural and sociocultural factors, in such diseases. The combination of genetic and environmental factors in many common diseases suggests that they may be preventable by avoidance, by the susceptible exposure to the environmental factors involved. A recent WHO (1983) working group noted that the use of genetic markers for chronic diseases of adult life holds great potential for risk prediction, prevention by avoidance of other relevant factors, and early treatment of the disease.

This IAHB conference on isolation and migration has two objects. The first is to draw together some of the outstanding research that has been carried out in this area, so that those carrying out medical and genetic surveys in different isolated and migrant populations, may perceive from what has been done what remains to be done. The ultimate goal of this research, on genetic susceptibility and environmental pathogenic factors of diet and life habit, is to improve the accuracy and effectiveness of current efforts to reduce the incidence of common diseases. This is the reason why our Association has decided to support the activities of the commission on medical anthropology and epidemiology, chaired by Professor Rudan. To help this endeavour forward, our discussions in Fukui, followed by the publication of the excellent papers presented, are the first steps. We are now convinced that the genetic and anthropological approach to the prevention of diseases will emerge as one of the dominant strategies for the enhancement and maintenance of health sciences in the 21st century, under the WHO theme of Health for All in 2000.

References

Collins, K. J. & Weiner, J. S. (1977). *Human adaptability: a history and compendium of research in the International Biological Programme*. London, Taylor & Francis.

Fujiki, N. (1981). Using family linkages to reconstruct an isolated Japanese villager's history. *World Conference on Records*. **2** (826), 1–31.

Fujiki, N., Nishigaki, I. & Mano, K. (1982). Genetic polymorphisms in isolated communities. *Japanese Journal of Human Genetics*, **27**, 121–130.

Ishimoto, G. & Yada, S. (1969). Frequency of red blood cell phosphoglucomutase phenotypes in the Japanese populations. *Human Heredity*, **19**, 198–202.

Komai, T. & Tanaka, K. (1972). Genetic studies on inbreeding in some Japanese populations. II. The Study of School Children in Shizuoka. *Japanese Journal of Human Genetics*, **17**, 114–148.

Mascie-Taylor, C. G. N. & Boyce, A. J. (1988). *Human mating patterns*. Cambridge: Cambridge University Press.

Matsumoto, H. (1966). *Serum Group*, pp. 5–37. Igaku Shoin.

Mourant, A. E., Kopec, A. C. & Domaniewska-Sobczak, K. (1976). *The distribution of human blood groups and other polymorphisms*, 2nd edn. Oxford University Press.

Mourant, A. E., Kopec, A. C. & Domaniewska-Sobczak, K. (1978). *Blood groups and diseases*. Oxford University Press.

Nakanaga, M. (1990). Population genetics study in an isolated community (IX) Miyamacho, with special reference on genetic susceptibility. *Japanese Journal of Const. Medicine*, **54**, 1–22.

Neel, J. V. & Schull, W. J. (1965). *The effects of inbreeding in Japanese children*. New York, Harper & Row.

Nei, M. & Imaizumi, Y. (1966). Genetic structure of human population. 1. Local differentiation of blood group gene frequency in Japan. *Heredity*, **21**, 9–35.

Omoto, K. & Harada, S. (1969). Polymorphisms of red cell acid phosphatase in several population groups in Japan. *Japanese Journal of Human Genetics*, **14**, 17–27.

Schull, W. J., Furusho, T., Yamamoto, M., Nagano, H. & Komatsu, I. (1970). The effect of parental consanguinity and inbreeding in Hirado, Japan (IV). Fertility and reproductive compensation. *Humangenetik*, **9**, 294–315.

Schull, W. J., Yanase, T. & Nemoto, H. (1962). Kuroshima: the impact of religion on an island's genetic heritage. *Human Biology*, **34**, 271–298.

Watanabe, S., Kondo, S. & Natsunaga, E. (1975). *Anthropological and genetic studies on the Japanese IBP synthesis* (II) (Human Adaptability). Tokyo: Tokyo University Press.

W.H.O. (1983). *Proposal for the multinational monitoring of trends and determinants in cardiovascular disease and protocol*. W.H.O./MNC 82.1.

Yamamoto, M., Fujiki, N., Nakanishi, K., Nakanishi, Y., Wada, T., Kanazawa, H., Nakai, T., Kondo, M., Hosokawa, K. & Masuda, M. (1974). Inbreeding coefficients in Arihara village, from pedigree study, phenotype and mating frequencies and correlation method. *Japanese Journal of Human Genetics*, **19**, 217–227.

2 *The distinction between primary and secondary isolates*

JAMES V. NEEL

In the English language the term isolate designates 'a relatively homogeneous population separated from related populations by geographic or biological or social factors or by the intervention of man' (Webster's Third New International Dictionary, 1971). This broad definition embraces a number of different genetic situations. Since, in any genetic study of an isolate, the expectation with respect to the distribution of allele frequencies and genetic disease differs according to the biological basis of the isolate, it is of some importance in genetics to distinguish between the various types of isolates. I suggest that from the genetic standpoint there are three principal types.

The first, the primary isolate, is typified by tribal populations of presumably very ancient origin which since they emerged as distinct entities have had relatively little biological exchange with other similar groups. In theory these populations should be as close to genetic equilibrium as any contemporary human populations although the stochastic process in these small subdivided groups ensures that equilibrium - if ever realised - is a fleeting phenomenon.

The second and third types of isolate, both of which may be termed secondary, come into being when a group for some reason detaches itself - or is detached - from a larger, usually national population which is a relatively recent amalgamation of smaller groups. The distinction between these two types depends simply on size. Thus, the second type of isolate to be delineated results when a relatively large group detaches itself from a still larger group and migrates into a new setting. The best examples of this would be the ethnic isolates found in many large cities - the Chinese-American group of San Francisco, the Korean population of Hiroshima, the Pakistani population of London. These ethnic isolates are sometimes referred to as 'minority' populations. The third type of isolate, by contrast is derived from a relatively small population sample, which then slowly expands, with very little recruitment from outside the group. An extreme example in Japan is the population of Hosojima Island.

The test of any classification is of course how well it accommodates the actual data. Let us consider the examples mentioned in this volume. That of Dr. Papiha refers to the Andamanese and Nicobarese which appear to be primarily tribal-type isolates. That of Dr. Garruto appears to be concerned with the third type of isolate, based on a small founding population. The presentation of Dr. Kondo and the mountain peoples referred to by Dr. Papiha appear to involve the second type of isolate, based on a relatively large founding population.

The genetic profiles of these three groups all differ. First let us consider the allele profile of a tribal isolate. Although no human population is ever in true genetic equilibrium, a long isolated, highly endogamous population approaches this situation, with an approximation to a balance between entry of new alleles into the population through mutation and loss through drift and selection. One of our salient findings in groups of this type, namely tribal Amerindians, has been the relatively high frequency of 'private polymorphisms', i.e., alleles restricted to a single tribe which, originating as mutational events, have through drift or selection achieved polymorphic frequencies (Neel, 1980). On the basis of a survey of some 13 Amerindian tribes with respect to electrophoretic variants at some 25 loci, we have estimated that approximately 1 in 40 loci supports such a private polymorphism. The frequency at the DNA level is of course much higher.

Second, let us consider the allele profile of an ethnic isolate, the second type of isolate, which is usually a relatively large subsample of an established, national population. But these established populations have within the past two to four thousand years been formed by the coalescence of tribes. This results in two changes in the allele structure of the population. First, since each of these tribes has contributed to the gene pool whatever unique alleles it possessed, the number of different rare variants encountered at a locus in a sample from one of these large ethnic isolates will be greater than in a sample of the same size from a tribal isolate. Second, through this admixture the private polymorphisms will be diluted to a lower gene frequency, now corresponding to rare variants. Populations such as this most definitely do not approach genetic equilibrium. Chakraborty *et al.* (1988) have recently provided a formal treatment of this phenomenon.

Now let us consider the allele profile of the third type of isolate, usually derived from a relatively recent but small sampling of a larger ethnic group, which then expands, sometimes dramatically. Here the allele profile is primarily determined by the 'luck of the draw', i.e., is a highly stochastic event. Because most humans carry several recessive genes with undesirable phenotypic effects, these populations are often characterised by particular inherited recessive diseases. Again, these are not equilibrium populations.

Isolates, because of their higher frequency of consanguineous marriage, have invited studies of inbreeding effects, especially in Japan. We come now to an important theoretical point. Because of these differences in allelic structure, the effects of inbreeding should be quite different in the three types of isolates. The relatively large (conglomerate, ethnic) isolate drawn from a national population should have a higher proportion of alleles of low frequency than the other two types of isolates. Since the magnitude of the inbreeding effect is a function of the rarity of the allele involved, the inbreeding effect, with reference to alleles conferring a phenotypic disadvantage, should be greater in large isolates drawn from national populations than in long-isolated tribal groups or isolates with a few founders drawn from a national, conglomerate population. The ethnic isolates should, in fact, exhibit the same inbreeding effect as the national population from which they were drawn.

Studies on inbreeding may be conducted both for their empirical value and for the insight they yield into the genetic dynamics of populations. In principle, the most appropriate populations to provide, through studies of inbreeding effects, an appraisal of the mutation-selection balance are the remaining relatively undisturbed tribal populations of the world. Unfortunately, the remoteness and the lack of any kind of written record render studies of inbreeding effects in these populations almost impossible. On the other hand, extensive studies have been carried out on the other two types of isolates.

Studies on type 3 isolates, arising from small groups which have become detached from larger civilised populations, are of principal value in defining the kinds of disease-producing alleles in the human genome. Dr. Yanase and Dr. Fujiki have studied such populations extensively. Our own experience with the population of Hosojima Island will serve as an example of this type of isolate (Ishikuni *et al.*, 1960). Hosojima is a small island in the Inland Sea which was apparently first settled by a fleeing samurai of the Matsumato clan and his wife, a samurai who had the misfortune to have been allied with the losing Hosokawa clan in the Battle of Funaokayama in Yamashiro-no-kuni in Eisho 8 (1511). Over the years, very few additional outsiders came to the island. At the time of our study, in 1959, there were 175 persons on the island. Twenty-nine of the 45 marriages on the island were demonstrably consanguineous. This resulted from the practice of marrying within the island if possible, in which case there was little choice but to marry a cousin. Of the 175 islanders, 11 (6.2%) exhibited a recessively inherited hereditary nerve deafness, often accompanied, depending on the age of onset, by mutism. Given the background of the island, there cannot be said to be an excess of consanguineous marriage over random expectation. Therefore, we can use Hardy-Weinberg equilibrium theory for an approximate estimate of the allele frequency. This is 0.25, the high frequency being yet another example of founder effect.

As an example of a study on a type 2 isolate, a large, ethnic isolate, we might consider a study undertaken on the island of Hirado in the late 1960s, again in collaboration with Dr. Yanase, Dr. Fujiki, Dr. Nakajima, and their colleagues (cf. Schull & Neel, 1972; Schull *et al.*, 1970a,b; Neel *et al.*, 1970a, b). At the time of these studies, it was a matter of some contention whether the outcome of consanguineous marriage suggested that the genetic variation exhibited by populations was predominantly maintained by mutation pressures or by heterozygote advantage. Morton *et al.* (1956) had argued that the ratio of the mortality predicted at full inbreeding ($F=1$) from the regression of indicator on degree of consanguinity, termed B, to the mortality in the outbred (control) population, termed A, provided a measure by which to answer the question. The uncertainties introduced in the extrapolation from inbreeding coefficients usually not exceeding 0.06 to a coefficient of 1.0 were obviously tremendous, as was the bias introduced by the fact that the intercept (A) included death from all causes, only a fraction of which was genetic. Nevertheless, based primarily on the data of Sutter and Tabah (1953) from France, which yielded a B/A ratio of 18.5, they concluded that the 'genetic load' revealed by inbreeding was primarily mutational in origin.

The Hirado study yielded a B/A ratio of 6.7. We suggested that this lower B/A ratio was most appropriately explained by segregation involving a preponderance of balanced genetic systems of various types, i.e., a segregational load. Other contemporary studies on urban populations in Hiroshima, Nagasaki, and Shizuoka (summary in Schull & Neel, 1965, 1972), had yielded similar B/A ratios. I note in this connection that in the simplest case, a balanced polymorphism leads to a B/A ratio of 2:1, but B/A ratios as high as 10:1 can readily be explained by a population structure in which the preponderance of the inbreeding effect results from a segregational load (Neel & Schull, 1962).

While these consanguinity studies were in progress in Japan, the interpretation of inbreeding results became more complicated. A number of investigators developed the concept of truncation selection as a more effective mode of elimination of deleterious genes than the independence of gene action implied in the Morton-Crow-Muller formulation (King, 1967; Milkman, 1967; Sved *et al.*, 1967). Other investigators explored the role of linkage in creating 'super' genes which would function as a unit (Wills *et al.*, 1970; Franklin & Lewontin, 1970). We now see, from this consideration of the agglomerate nature of most of the civilised populations on which studies of inbreeding effects have been performed, a further complication in the interpretation of inbreeding results, namely, how far this type of population departs from the equilibrium conditions which must be postulated to interpret inbreeding results. In principle, allowance for this 'agglomeration factor' would only lower the already low B/A ratios observed in the Japanese studies. The effect of inbreeding in an equilibrium population should thus be substantially less than observed in Japan. Why, in the face of mutation rates of the order of 10^{-5}/locus/generation, or 10^{-8}/nucleotide/generation, are inbreeding effects not larger? This remains an important question in human genetics.

References

Chakraborty, R., Smouse, P. E. & Neel, J. V. (1988). Population amalgamation and genetic variation: observations on artificially agglomerated tribal populations of Central and South America. *American Journal of Human Genetics*, **43**, 709–725.

Franklin, I. & Lewontin, R. C. (1970). Is the gene the unit of selection? *Genetics*, **65**, 707–734.

Ishikuni, N., Nemoto, H., Neel, J. V., Drew, A. L., Yanase, T. & Matsumoto, Y. S. (1960). Hosojima. *American Journal of Human Genetics*, **12**, 67–75.

King, J. L. (1967). Continuously distributed factors affecting fitness. *Genetics*, **55**, 483–492.

Milkman, R. D. (1967). Heterosis as a major cause of heterozygosity in nature. *Genetics*, **55**, 493–495.

Morton, N. E., Crow, J. F. & Muller, H. J. (1956), An estimate of the nutritional damage in man from data on consanguineous marriages. *Proceedings of the National Academy of Sciences, USA*, **42**, 855–863.

Neel, J. V. (1980). Isolates and private polymorphisms. In *Population Structure and Disorders*, ed. A. Erikkson, pp. 173–193. London: Academic Press.

Neel, J. V. & Schull, W. J. (1962). The effect of inbreeding on mortality and morbidity in two Japanese cities. *Proceedings of the National Academy of Sciences, USA*, **48**, 573–582.

Neel, J. V., Schull, W. J., Kimura, T., Tanigawa, Y., Yamamoto, M. & Nakajima, A. (1970a). The effects of parental consanguinity and inbreeding in Hirado, Japan. III. Vision and hearing. *Human Heredity*, **20**, 129–155.

Neel, J. V., Schull, W. J., Yamamoto, M., Uchida, S., Yanase, T. & Fujiki, N. (1970b). The effects of parental consanguinity and inbreeding in Hirado, Japan. II. Physical development, tapping rate, blood pressure, intelligence quotient, and school performance. *American Journal of Human Genetics*, **22**, 263–286.

Schull, W. J., Furusho, T., Yamamoto, M., Nagano, H. & Komatsu, I. (1970a). The effects of parental consanguinity and inbreeding in Hirado, Japan. IV. Fertility and reproductive compensation. *Humangenetik*, **9**, 294–315.

Schull, W. J., Nagano, H., Yamamoto, M. & Komatsu, I. (1970b). The effects of parental consanguinity and inbreeding in Hirado, Japan. I. Stillbirths and prereproductive mortality. *American Journal of Human Genetics*, **22**, 239–262.

Schull, W. J. & Neel, J. V. (1965). *The Effects of Inbreeding on Japanese Children*, pp. xii + 419. New York: Harper & Row.

Schull, W. J. & Neel, J. V. (1972). The effects of parental consanguinity and inbreeding in Hirado, Japan. V. Summary and interpretation. *American Journal of Human Genetics*, **24**, 425–453.

Sutter, J. & Tabah, L. (1953). Structure de la mortalite dans les familles consanguines. *Population*, **8**, 511–526.

Sved, J. A., Reed, T. E. & Bodmer, W. F. (1967). The number of balanced polymorphisms that can be maintained in a natural population. *Genetics*, **55**, 469–481.

Wills, C., Crenshaw, J. & Vitale, J. (1970). A computer model allowing maintenance of large amounts of genetic variability in Mendelian populations. I. Assumptions and results for large populations. *Genetics*, **64**, 107–123.

3 *Time trends in the break-up of isolates*

TOSHIYUKI YANASE

In many areas of the world this century, isolates have tended to break up, and particularly since World War II. Sutter and Tabah (1954) were among the first to draw attention to the importance of this process, which they termed 'éclatement' (bursting) and which they attempted to measure by changes in demographic variables, e.g. endogamy, consanguineous unions. In rural areas of Japan, traditional ways of selecting marriage partners still prevail, but the tendencies to endogamy in various populations have declined year by year.

In 1966 were investigated time trends in isolates based on the frequencies of consanguineous marriages in four different communities in Japan (Yanase, 1966). Because of the complexity of the network of biological relationships in these communities (Table 3.1), analysis was restricted to only those marriages where there was consanguinity corresponding to a coefficient of inbreeding for the offspring of 1/16. It became apparent that the higher the degree of isolation the more rapid was the break-up of the isolate (Figure 3.1).

Such trends in consanguinity may be attributed to demographic variables quite apart from migration, such as a decrease in the mean number of children born, or in the variance (Table 3.2), as well as to a reduced non-random tendency in recent marital practices. In addition, there are significant differences in reproductive performance between native and non-native spouses in these communities.

Human isolates can be divided into four categories:

1. Offshoot populations in areas geographically distant from the main population.
2. Communities isolated on account of a particular factor, for example, some 2,300 inhabitants on an island in Nagasaki Prefecture clearly constituted two subgroups: one Buddhist and the other Roman Catholic. Very few marriages were contracted between persons from these two religious groups during the past hundred years (Schull *et al.*, 1962).

Table 3.1. *Distribution of coefficients of inbreeding of 396 consanguineous matings in community Hs*

Coefficient of inbreeding	*Number of cases (%)*
.18700	1
.12500 uncle-niece	16 (4%)
.10938	2
.09766	1
.09375	3
.08984	1
.08534	2
.08203	2
.07130	25 (6%)
.07031	1
.06731	1
.06641	3
.06348	1
.06250 first cousins	195 (49%)
.05566	1
.05078	1
.04688	2
.03906	1
.03516	2
.03203	1
.03125 first cousins once removed	59 (15%)
.02344	6
.01953	4
.01563 second cousins	47 (12%)
.01172	1
.00781 second cousins once removed	10 (3%)
.00391 third cousins	5 (1%)
.00195 third cousins once removed	2

The biological relationships were identified by the koseki registers and interview with the family (Yanase, 1962; Yanase *et al.*, 1973).

3. Unique ethnic communities, such as the Ainu who inhabit the northernmost part of Japan.
4. Primitive populations of hunter-gatherers still living under conditions comparable with those during prehistoric times.

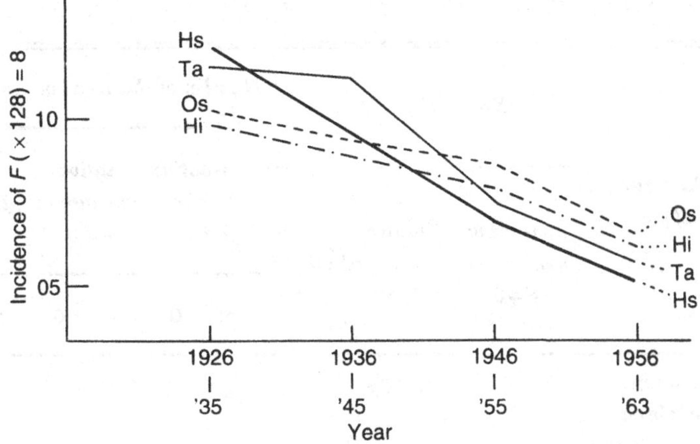

Figure 3.1. Time trends in isolates (1926-1963). Four different communities designated as Hs, Ta, Os, and Hi (Yanase, 1966).

The rapidity of break-up will vary according to the type of isolate, the resilience of the culture, and the nature and intensity of pressures from outside. The biological effects of that break-up will also therefore vary.

Biological effects of break-up

Isolation obviously affects breeding structures, migration patterns, and particularly genetic distance, its theory and actual applications, in other words factors affecting the evolutionary dynamics of a population. But also there are to be considered the biological effects of isolation and its breakup, for example on differentiation of population subgroups, morbidity, mortality, fertility, etc.

In 1951, during investigations of a mountainous population of about 7,000 inhabitants in Miyazaki Prefecture, subgroups of kindreds living in a deep valley showed 16 different major congenital abnormalities. There were 5 kindreds with congenital nystagmus, 4 with congenital deafness, 3 with idiopathic epilepsy, one with ectodermal dysplasia of unknown etiology, etc. The total affected individuals numbered 89 (Yanase, 1951, 1964). Each of the units had no recent blood relationship with the others, as determined from tracing ancestry over the previous 6-7 generations. Such an unusually high density of specific major abnormalities is probably due partly to a founder effect, and also partly to a residuum of mentally and physically handicapped individuals who did not join those who emigrated from the isolates

Table 3.2. *Number of children born for the years 1916-1955*

Marriage year of parents	*Number of sibships*			*Number of children born (m and σ)*					
	non-native non-inbred F=0	*native non-inbred F=0*	*native inbred F>0*	*non-native non-inbred F=0*		*native non-inbred F=0*		*native inbred F>0*	
				m	σ	*m*	σ	*m*	σ
Hs community (an inland population)									
1916-25	38	67	44	5.7	13.3	6.3	11.1	6.8	8.1
1926-35	59	100	35	5.3	6.7	5.9	5.7	6.1	6.2
1936-45	91	106	28	3.8	3.0	3.8	3.0	3.8	3.0
1946-55	122	159	55	2.9	1.9	3.3	1.4	3.3	1.4
Os community (an island population)									
1916-25	8	29	11	4.9	5.9	7.4	11.2	6.7	8.7
1926-35	24	37	21	3.9	9.1	5.8	5.0	5.6	4.0
1936-45	31	22	13	3.4	4.4	3.8	2.7	4.6	4.7
1946-55	39	68	31	2.9	1.3	3.1	1.5	3.1	2.5

Individuals are subdivided in accordance with degree of endogamy and inbreeding coefficient (F) (Yanase, 1964).

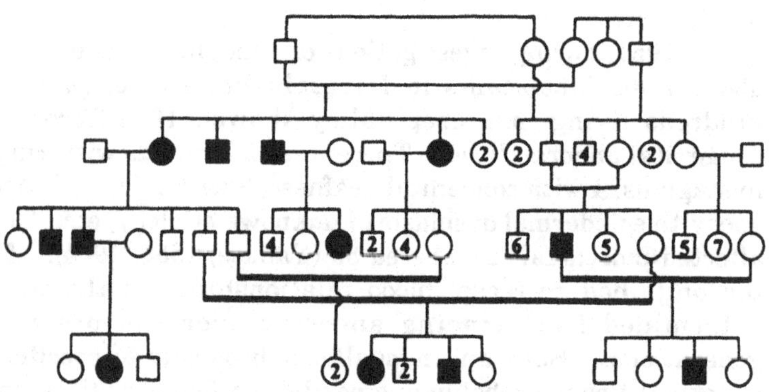

Figure 3.2. A family with alkaptonuria discovered in a rural community with high endogamy (Abe *et al.*, 1960).

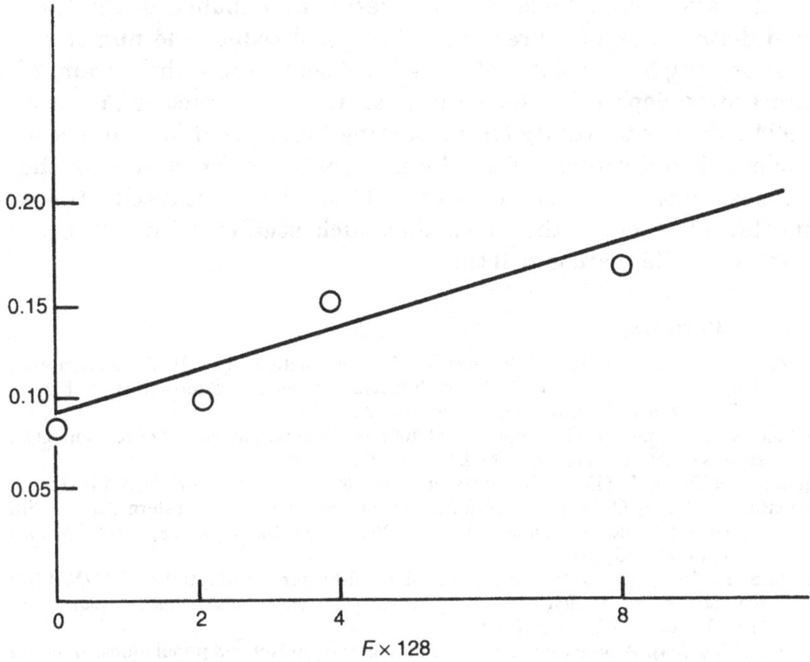

Figure 3.3. Regression of early mortality on the degree of inbreeding in an inland community (Yanase, 1970).

into urban and city populations. Whatever the factors responsible for them, such high densitites will no longer occur as the intensity of isolation fades,

Abe *et al.*, (1960) observed a family with alkaptonuria in a rural community with high endogamy (Figure 3.2). This metabolic error is an autosomal recessive condition and is not thought to be genetically heterogeneous. Yet the pedigree suggests a pseudodominant mode of inheritance. This can only result from matings between individuals homozygous and heterozygous for a rare recessive allele, and is likely to be due to incompleteness of the pedigree as drawn, which shows as unrelated spouses who were in fact related and who had received the gene from a remote common ancestor. Such an extremely rare case would hardly be detectable in populations other than isolates.

Figure 3.3 shows a linear regression of early mortality on the degree of inbreeding (F) in an inland community of a relatively large population located near urban areas of Fukuoka Prefecture. It is feasible to estimate lethal equivalents even in isolates if sufficient samples for analysis can be obtained.

Break-up of isolates then will result in a change in the level and distribution of inbreeding. This will reduce the number of manifesting homozygotes of recessive disorders and the amount of inbreeding depression shown in quantitative characters (Yanase, 1984). The opportunity for elucidating the mode of inheritance of undescribed disorders of the former type, and for analysing the genetic contribution to the latter will therefore be reduced. It is a matter of urgency therefore that such studies of isolates be pursued while there is still time.

References

Abe, Y., Oshima, N., Hanakita, R., Amako, T. & Hirohata, R. (1960). Thirteen cases of alkaptonuria from one family tree with special reference to osteo-arthrosis alkaptonuria. *Journal of Bone & Joint Surgery*, **42 – A**, 817–831.

Schull, W. J., Yanase, T. & Nemoto, H. (1962). Kuroshima: The impact of religion on an island's genetic heritage. *Human Biology*, **34**, 271–298.

Sutter, J. & Tabah, L. (1954). The break-up of isolates. *Eugenics Quarterly*, **3**, 148–154.

Yanase, T. (1951), Genetic studies on highly inbred villagers in western Japan, with special reference to Shiiba, Miyazaki Prefecture (in Japanese). *Acta Medica Fukuoka*, **21**, 183–209.

Yanase, T. (1962). Use of the Japanese family register for genetic studies. In *UN/WHO Seminar on Use of vital and health statistics for genetic and radiation studies*, pp. 119–133. New York: United Nations.

Yanase, T. (1964). A note on the patterns of migration in isolated populations. *Japanese Journal of Human Genetics*, **9**, 136–152.

Yanase, T. (1966). A study of isolated populations (The Japan Society of Human Genetics Award Lecture). *Japanese Journal of Human Genetics*, **11**, 125–161.

Yanase. T. (1970). Clinical genetics in internal medicine – Genetic analyses of morbidity (in Japanese). *Japanese Journal of Internal Medicine*, **59**, 917–935.

Yanase, T. (1984). Detrimental equivalents in man, with special reference to prospective views on change in prevalence of SLE, a multifactorial disease, in the next generation – Presidential lecture (in Japanese). *Japanese Journal of Internal Medicine*, **73**, 1285–1297.

Yanase, T., Fujiki, N., Handa, Y., Yamaguchi, M., Kishimoto, K., Furusho, T. & Tanaka, K. (1973). Studies of isolated populations. *Japanese Journal of Human Genetics*, **17**, 332–367.

4 *Factors influencing the frequency of consanguineous marriages in Japan*

YOKO IMAIZUMI

Introduction

In a nationwide survey, the rate of total consanguineous marriages in Japan decreased from 16% to 3% in the period of 25 years from 1947 (Imaizumi *et al.*, 1975). This tendency continued to 1983, as shown by Imaizumi (1986a) in six sample areas representative of all Japan (Figure 4.1). According to Shinozaki (1961), the rate of consanguinity in Japan for the marriage years from 1912 to 1925 was 22.4%, it then slowly decreased to about half that level (12.3%) for the period 1941 to 1945, and after the Second World War a rapid decrease occurred. The inbreeding coefficient (F) in isolated populations in Japan decreased during the period from 1926 to 1963 (Yanase, 1966), though there remained appreciable variation. From the nationwide survey data of Imaizumi *et al.* (1975) the mean F value in rural areas in Japan (0.0029) was twice that in urban areas (0.0014), and out of eight districts it was highest in Kyushu (0.0034) and lowest in Hokkaido (0.0009). Similar decreases with time have occurred elsewhere, e.g. in Spain the inbreeding coefficient (F) remained constant during the period from 1900 to 1959 and has decreased consistently over the last two decades (Calderon, 1989).

The decline of the consanguinity rate is usually attributed to higher mobility of the population and therefore a wider choice of possible spouse. This paper investigates the effects of marital distance (between birthplaces), religion, socioeconomic factors, marriage form, and opportunity of meeting on the rate of consanguineous marriages in Japan, and also reports the stated reasons for consanguineous marriages: it summarises results reported elsewhere (Imaizumi, 1986a,b,c, 1987, 1988).

Materials and methods

This study utilises data from the Demographic Survey of Japanese Marriages conducted on September 1st 1983 by the Institute of

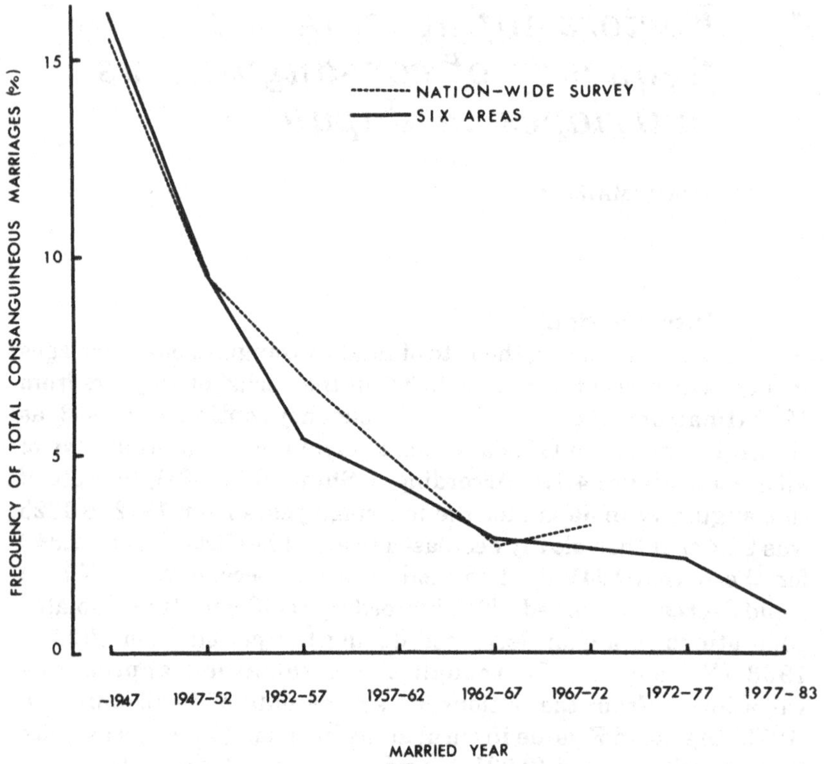

Figure 4.1. Frequency of consanguineous marriages according to marriage year (Imaizumi *et al.*, 1975; Imaizumi, 1986a).

Population Problems, Ministry of Health and Welfare. For this survey six areas were selected as representing different parts of Japan: Asahikawa City area (Hokkaido), Tagajo-Shi (Miyagi Prefecture), Minobu-Cho (Yamanashi Prefecture), Okazaki-Shi (Aichi Prefecture), Kawanishi-Shi (Hyogo Prefecture) and Fukue-Shi (Nagasaki Prefecture). The survey was limited to married couples under the age of 65 years. The number of couples sampled in each area was 1,600, so the total number of couples examined was 9,600. Questionnaires were distributed to 9,600 couples, and only 375 (3.9%) were not returned.

Consanguineous marriages more remote than between second cousins once removed and between those of unknown relationships were regarded as non-consanguineous, i.e. the inbreeding coefficients of their offspring were regarded as zero in the computation of the mean F value.

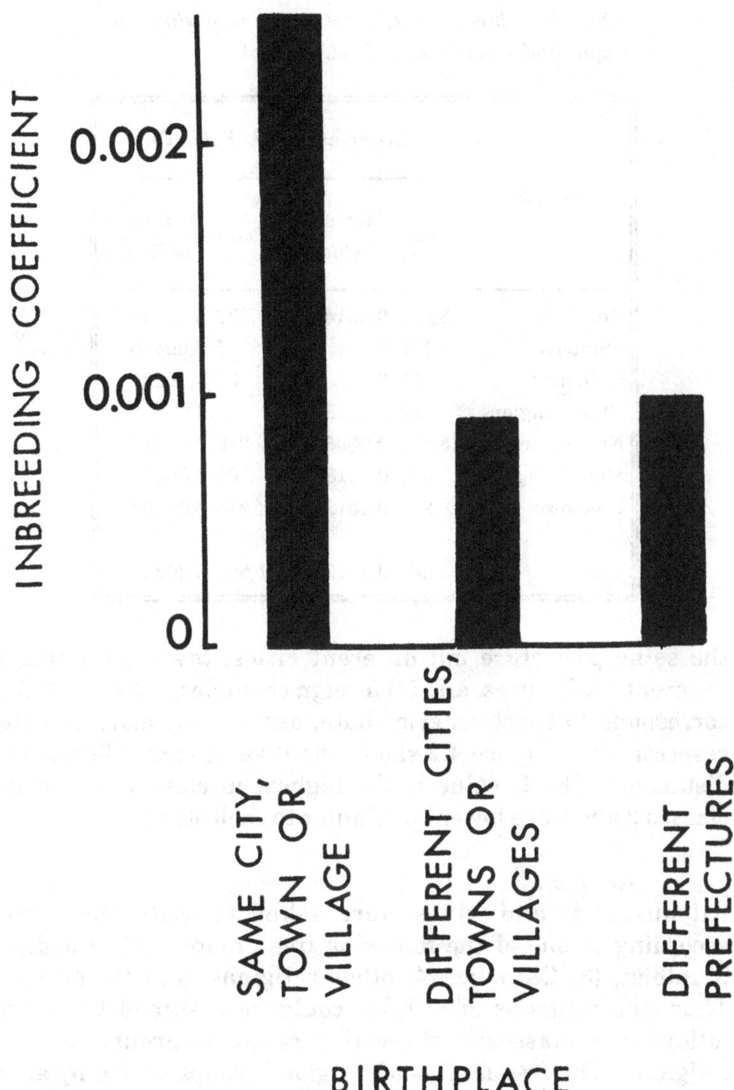

Figure 4.2. **Inbreeding coefficient according to marital distance (Imaizumi, 1986c).**

Results

Marital distance

Unions were classified by birthplaces of spouses as follows: spouses born in (1) the same city, town, or village (endogamy), (2)

Table 4.1. *Inbreeding coefficient according to religion of couples and area* (From Imaizumi, 1986b)

Religion	*Five areas*		*Fukue City*	
	Total	*Inbreeding coefficient*	*Total*	*Inbreeding coefficient*
Buddhist	3,005	0.00142	972	0.00327
Shintoist	119	0	35	0.00446
Catholic	21	0	58	0.00121
Other religions	52	0.00150	8	0
No religion	3,134	0.00069	210	0.00216
Mixed religion	397	0.00138	60	0.00104
Unknown	939	0.00123	215	0.00102
Total	7,667	0.00107	1,558	0.00266

the same prefecture but different cities, towns or villages, (3) different prefectures, and (4) foreign countries. Classes 1, 2, and 3 correspond to short, intermediate, and remote marital distances, respectively. Figure 4.2 shows the F values according to marital distance. The F value is the highest in class 1, and values in classes 2 and 3 are lower but similar to each other.

Religion

All husbands and wives were asked to state their religion according to one of the following five groups: (1) Buddhist, (2) Shintoist, (3) Catholic, (4) other religions, and (5) no religion. Thus, the religions of couples could be assigned to one of the following 6 classes: these five religious groups and mixed-religion. The frequencies of religious groups varied by area, but Fukue City was also quite different from the other five areas (Imaizumi, 1986b). Computation of the F values by religious groups (Table 4.1) shows the mean inbreeding coefficient in Fukue City to be 2.5 times that in the other five areas combined (respectively 0.0027 and 0.0011). In the five other areas, the F value is similar for all religions except Shintoist and Catholic where it is 0, and intermediate (0.0007) for those of 'no religion'. By contrast in Fukue City, the F value is the highest for Shintoists (0.0045). These are followed by Buddhists (0.0033) who are the group mainly reponsible for the higher overall mean and indeed

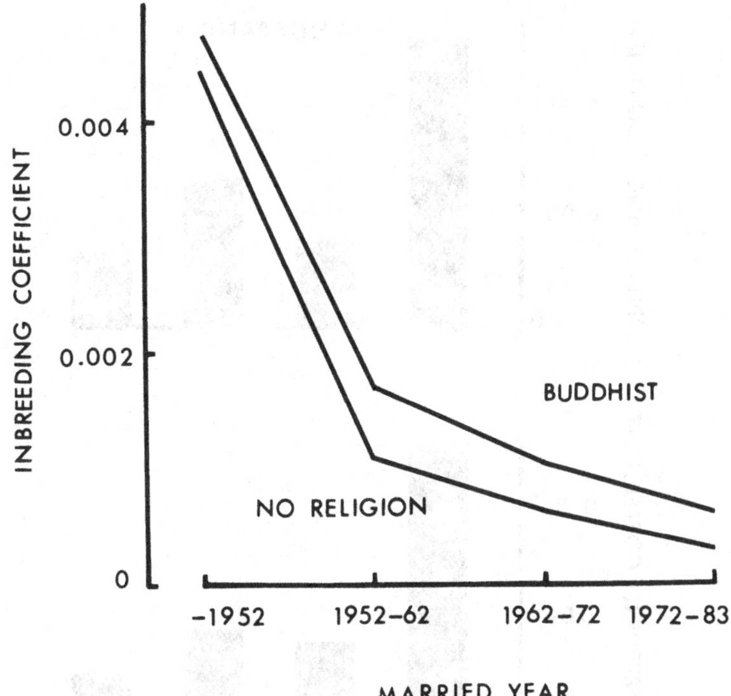

Figure 4.3. Inbreeding coefficient according to marriage year and religion of couples (Imaizumi, 1986b).

their mean F value is twice that for the 'no religion' group in the five other areas, 1.5 times in Fukue City, and 2.4 times overall. The F values for Buddhists, the irreligious, and mixed-religion groups decrease with marriage year in the five areas; and for the former two groups in Fukue City. When all areas are combined, the F value is higher in Buddhists than in the 'no religion' group in each marriage year (Figure 4.3).

Education and occupation

The type of school last attended was divided into five classes: (1) junior high school, (2) senior high school, (3) junior college or higher professional school, (4) college, university or graduate course, and (5) others. The mean F value according to the type of last school attended for husband and wife (Figure 4.4) shows a diminution with increasing education, being highest in class 1 for husbands (0.0025) and wives (0.0024), and the lowest in class 4 for husbands (0.0006) and wives (0.0005). The F values for husbands of class 1 are 4 times those for class 4, and for wives 5 times.

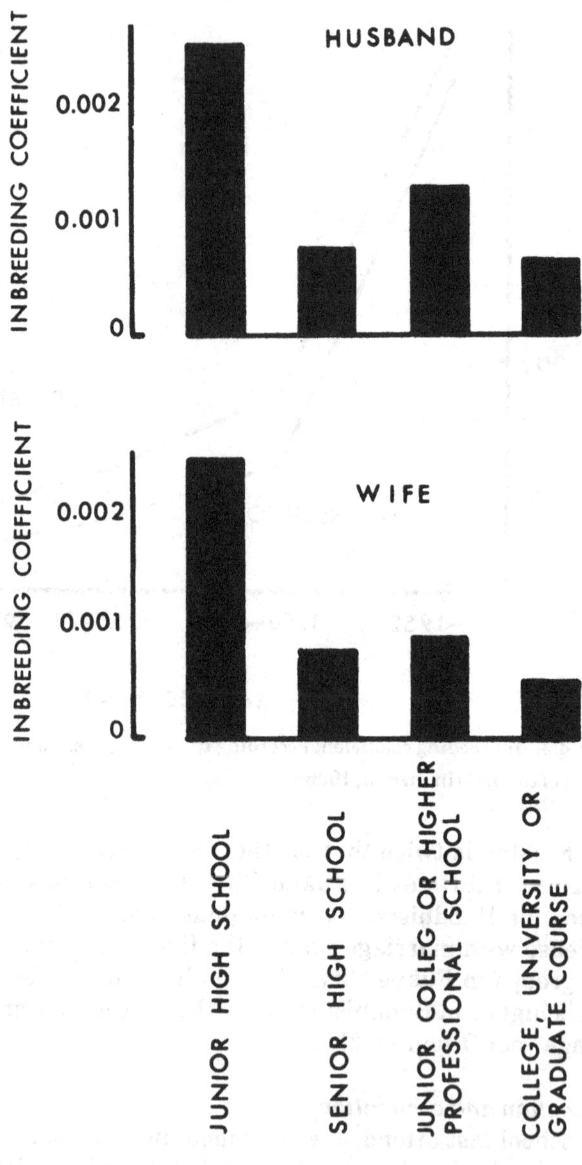

Figure 4.4. Inbreeding coefficient according to the type of last school attended for husband and wife (from Imaizumi, 1986b).

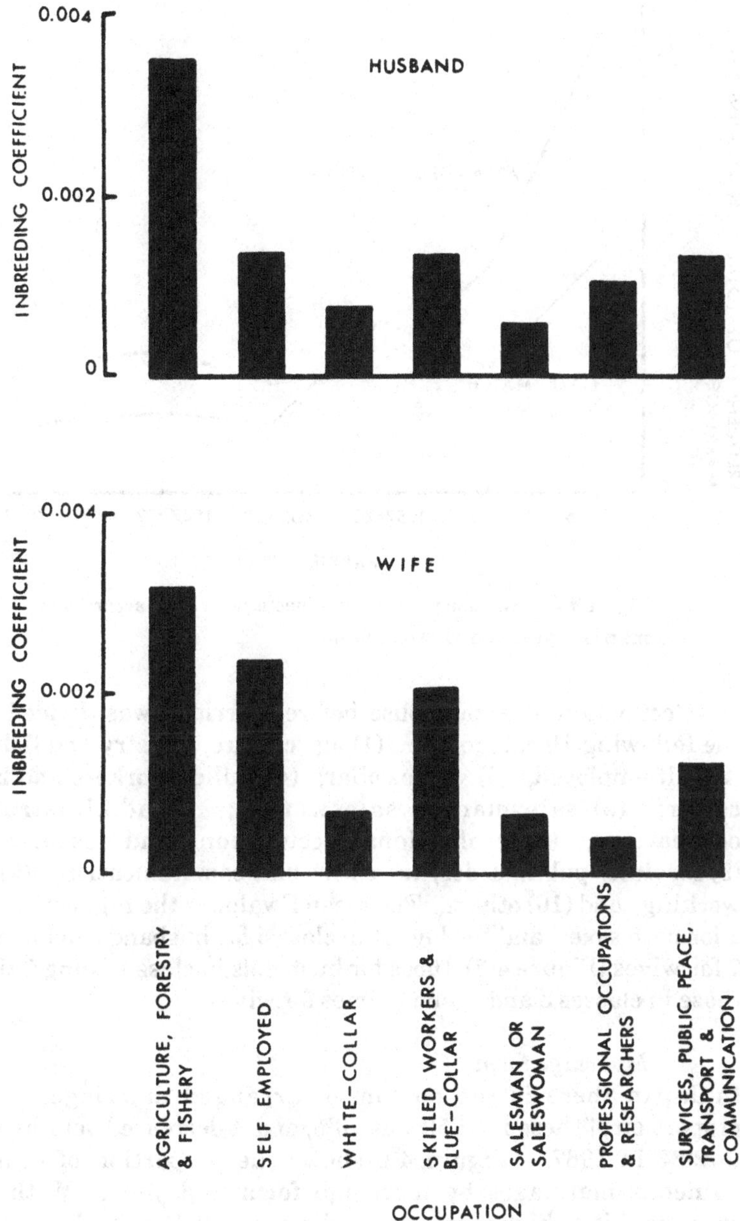

Figure 4.5. Inbreeding coefficient according to the occupation of husband and wife (Imaizumi, 1986b).

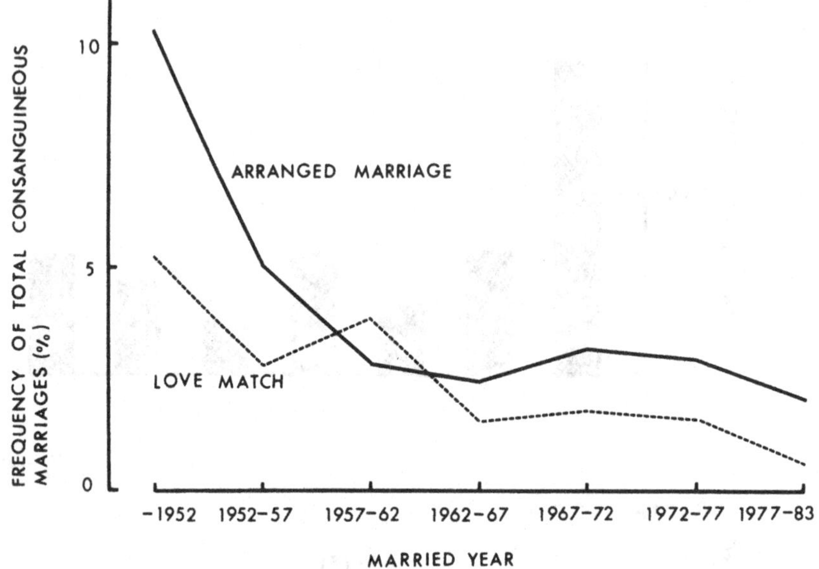

Figure 4.6. Frequency of consanguineous marriages according to marriage year and marriage form.

Occupation of each spouse before marriage was divided into the following 10 categories: (1) agriculture, forestry and fishery; (2) self-employed; (3) white-collar; (4) skilled workers and blue-collar; (5) salesman or saleswoman; (6) administrative occupations; (7) professional occupations and researchers; (8) services, public order, transport and communication; (9) not working; and (10) others. The mean F value is the highest in class 1 for both sexes, and the lowest in class 5 for husbands and in class 7 for wives (Figure 4.5), those for husbands in class 1 being 6 times those in classes 5 and 7, and 8 times for wives.

Marriage form

In Japan there is the custom of arranged marriage. The proportion of these in 1950 was 70%, and it decreased year by year to 20% in 1987. Figure 4.6 shows the proportion of consanguineous marriages by marriage form and year. With one exception, it is higher in arranged marriages than in love match marriages in each marriage year.

The opportunity of encounter

How the couple had met was considered by classifying the information as follows: (1) arranged meeting, (2) friendship since

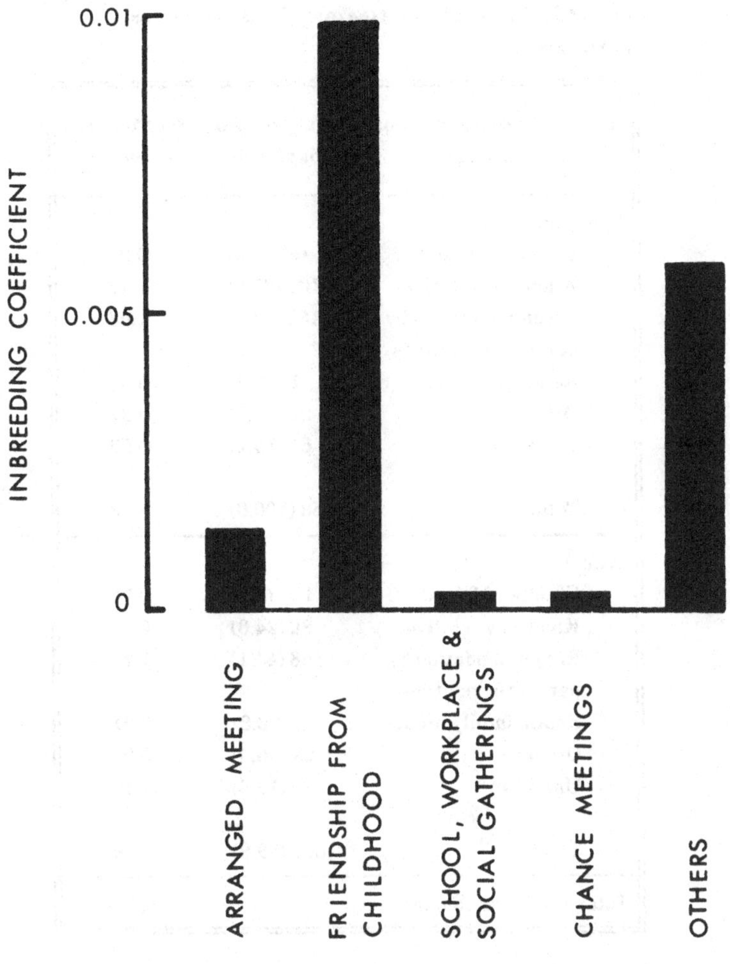

Figure 4.7. Inbreeding coefficient according to the opportunities of encounter (Imaizumi, 1986c).

childhood, (3) school, workplace and social gatherings, (4) chance meetings, and (5) others. The mean F value is the highest in class 2 (0.00986) and the lowest in class 4 (0.00027). The F value in class 2 is over 30 times that in classes 3 and 4, and 7 times that in class 1 (Figure 4.7). The F values decrease with the year of marriage in classes 1 and 2, but not in classes 3 and 4. It is higher in class 2 than in classes 1 and 3 in each year of marriage.

Table 4.2. *Reasons for and the rate of consanguineous marriages*

Reason for consanguineous marriage	*No. of related couples (%)*	*% of total no. of couples*
Husband		
Childhood friend	19 (5.3)	0.21
Known by relatives	107 (29.9)	1.16
Recommendation by parents or relatives	155 (43.3)	1.68
Retain family property	1 (0.3)	0.01
Others	22 (6.1)	0.24
Unknown	54 (15.1)	0.59
Total	358 (100.0)	3.88
Wife		
Childhood friend	19 (5.3)	0.21
Known by relatives	86 (24.0)	0.93
Recommendation by parents or relatives	158 (44.1)	1.71
Retain family property	1 (0.3)	0.01
Others	23 (6.4)	0.25
Unknown	71 (19.8)	0.77
Total	358 (99.9)	3.88
Total number of couples		9,225

Reasons for consanguineous marriages

The husbands and wives in the 358 related couples were asked to select one of the following five reasons for their choice of a consanguineous spouse: (1) childhood friends, (2) known by relatives, (3) recommendation by parents or relatives, (4) to retain family property, and (5) others. Recommendation by parents or relatives is the most frequent reason in both husbands (43%) and wives (44%), followed by knowledge by relatives, both being greater than the existence of a childhood friendship (Table 4.2).

Discussion

According to Imaizumi (1977, 1978), the mean F value decreased and the mean marital distance increased with more recent year of marriage in Gyoda and Hasuda cities, Saitama Prefecture and Kanoya City, Kagoshima Prefecture. In the present study, the proportion at the smallest marital distance, endogamy, was 40% in the oldest year group and decreased gradually to 27% in the most recent marriage year group. The corresponding F values were 0.0045 and 0.00017, and as expected there was a high and significant ($p<0.05$) correlation coefficient (0.95) between the F value and the endogamy rate.

Schull *et al.* (1962,1968), and Yoshikawa (1977) studied the consanguinity by religion in Nagasaki Prefecture, where the rate of consanguineous marriages was highest in Shintoists followed by Buddhists, and lowest in Catholics. In the present study the same tendency is shown: the F value in Fukue City (in the same prefecture) is also the highest in Shintoists, followed by Buddhists, 'no religion' and Catholics. Elsewhere, in South India (Devi, 1982) the F value was the highest in Hindus (0.0267), followed by Muslims (0.0131), and the lowest in Christians (0.0118), and in Beirut, Lebanon (Khlat, 1988) the rate of consanguineous marriage was higher in Muslims (29.6%) than Christians (16.5%). Thus religious variation appears to be a general factor affecting the rate of consanguinity.

Schull and Neel (1965) and Komai and Tanaka (1972) examined the consanguinity of parents of school children and, since their categories of parental education and occupation are equivalent to those of education and occupation of husbands or wives in the present study, the results can be compared. The tendencies as to socioeconomic factors are similar and inbreeding levels in Japan are still associated with education and occupation.

According to Ichiba (1953), that prospective mates knew each other well (43%) was the most frequent reason for consanguineous unions, followed by geographical and economic factors (11.5%). Although strict comparison with Ichiba's data cannot be made since the categories of reason do not coincide exactly, the reason of 'knowledge by relatives' in the present study and of 'prospective mates knew each other well' in Ichiba's are quite similar in meaning, so that the lower figures (30% for husbands and 24% for wives) in the present study suggest a modest diminution in the importance of this reason.

Conclusion

This study shows the continuing existence of consanguineous unions as an important component in the marriage pattern in Japan, though the diminution in frequency noted in earlier years continues. There are obviously many factors affecting the

occurrence of consanguineous unions. These factors include marital distance, religion, education and occupation, the custom of arranged marriages, and the opportunities that exist for young people to meet. The reasons stated for consanguineous unions however show that even though there has been a diminution in the proportion of arranged marriages, recommendation of spouse by other family members still remains an important factor.

References

Calderon, R. (1989). Consanguinity in the Archbishopric of Toledo, Spain, 1900–79. I. Types of consanguineous mating in relation to premarital migration and its effects on inbreeding levels. *Journal of Biosocial Science*, **21**, 253–266.

Devi, A. R. R. (1982). Inbreeding in the State of Karnataka, South India. *Human Heredity*, **32**, 8–10.

Ichiba, M. (1953). Studies on the children of consanguineous marriages. I. Frequency and causes of consanguineous marriage (in Japanese). *Nihon Ika Daigaku Zasshi*, **20**, 798.

Imaizumi, Y. (1977). A demographic approach to population structure in Gyoda and Hasuda, Japan. *Human Heredity*, **27**, 318–327.

Imaizumi, Y. (1978). Population structure in Kanoya Population, Japan. *Human Heredity*, **28**, 7–18.

Imaizumi, Y. (1986a). A recent survey of consanguineous marriages in Japan. *Clinical Genetics*, **30**, 230.

Imaizumi, Y. (1986b). A recent survey of consanguineous marriages in Japan: religion and socioeconomic class effects on consanguineous marriages. *Annals of Human Biology*, **13**, 317–330.

Imaizumi, Y. (1986c). Factors influencing the frequency of consanguineous marriages in Japan: marital distance and opportunity of encounter. *Human Heredity*, **36**, 304–309.

Imaizumi, Y. (1987). Reasons for consanguineous marriages in Japan. *Journal of Biosocial Science*, **19**, 97–106.

Imaizumi, Y. (1988). Parental consanguinity in two generations in Japan. *Journal of Biosocial Science*, **20**, 235–243.

Imaizumi, Y., Shinozaki, N. & Aoki, H. (1975). Inbreeding in Japan: results of a nation-wide study. *Japanese Journal of Human Genetics*, **20**, 91–107.

Khlat, M. (1988). Social correlates of consanguineous marriages in Beirut: a population-based study. *Human Biology*, **60**, 541–548.

Komai, T. & Tanaka, K. (1972). Genetic studies on inbreeding in some Japanese populations. II. The study of school children in Shizuoka: History, frequencies of consanguineous marriages and their subtypes, and comparability in socio-economic status among consanguinity classes. *Japanese Journal of Human Genetics*, **17**, 114–148.

Schull, W. J., Komatsu, I., Nagano, H. & Yamamoto, M. (1968). Hirado: Temporal trends in inbreeding and fertility. *Proceedings of the National Academy of Sciences* **59**, 671–679.

Schull, W. J. & Neel, J. V. (1965). *The Effect of Inbreeding on Japanese Children*. New York: Harper and Row.

Schull, W. J., Yanase, T. & Nemoto, H. (1962). Kuroshima: The impact of religion on an island's genetic heritage. *Human Biology*, **34**, 271–298.

Shinozaki, N. (1961). A somatological and genealogical study of the inhabitants of a consanguineous community in Japan. (in Japanese). *Journal of Population Problems*, **83**, 31–70.

Yanase, T. (1966). A study of isolated population. *Japanese Journal of Human Genetics*, **11**, 125–161.

Hoshikawa, I. (1977). Genetic studies on human populations of Goto-Islands II. Analysis of consanguineous marriages (in Japanese). *Nagasaki Igakkai Zasshi*, **52**, 300–316.

5 *Break-up of isolates*

FRIEDRICH VOGEL

At present mankind is experiencing a formidable demographic revolution. World population is increasing fast, mortality before and during reproductive age is being reduced, and traditional patterns of marriage are disappearing. One aspect of this demographic revolution of particular interest in population genetics is the break-up of isolates.

For millions of years, our predecessors lived in small population groups, often consisting of fewer than one hundred individuals. Each of these groups was to some extent isolated from others and usually marriage occurred within the group, sometimes between subgroups. As a consequence, their genetic composition changed relatively quickly, depending on chance fluctuations of gene frequencies, and to a certain degree under the influence of changing patterns of natural selection.

Effects of isolation

Whereas in an infinitely large, randomly mating population the frequency (q) of a given allele depends on its mutation rate and selective advantage or disadvantage, its frequency in a small population is greatly affected by chance - provided only that its selective disadvantage is not too strong. Figure 5.1 shows the distribution of the frequency of an allele (initial mean $q=0.5$) at a diallelic locus in groups of populations depending on their effective sizes largely determined by the number of couples breeding. Mutation rates and back mutation rates are assumed to be identical. With small population size (a), many populations will be homozygous for one of the two alleles ($q=0$ or 1) whereas, for very large populations, q clusters around 0.50. An allele may become homozygous (=fixed) in a small population by chance fluctuation in frequency even against a certain selective disadvantage (Figures 5.2 and 5.3).

In medical genetics, the effects of such chance fluctuation have become of practical importance in the special case of rare alleles leading to hereditary diseases - especially, but not exclusively, those showing an autosomal recessive mode of inheritance.

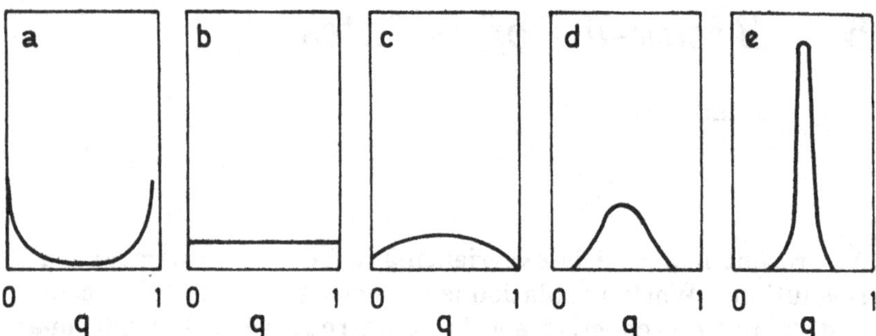

Figure 5.1. Distribution of gene frequency q in small populations in relation to the effective breeding population size N. Mutation rates μ and back-mutation rates υ are assumed to be identical. (a) $N\mu$ very small; (b) $4N\mu = 4N\upsilon = 1$; (c) $4N\mu = 4N\upsilon = 1.5$; (d) $4N\mu = 4N\upsilon = 10$; (e) $4N\mu = 4N\nu = 20$. With small population size (a), many populations will be homozygous for one of the two alleles ($q = 0$ or $q = 1$); with very large population size (e), q clusters around 0.5. (Li, 1955).

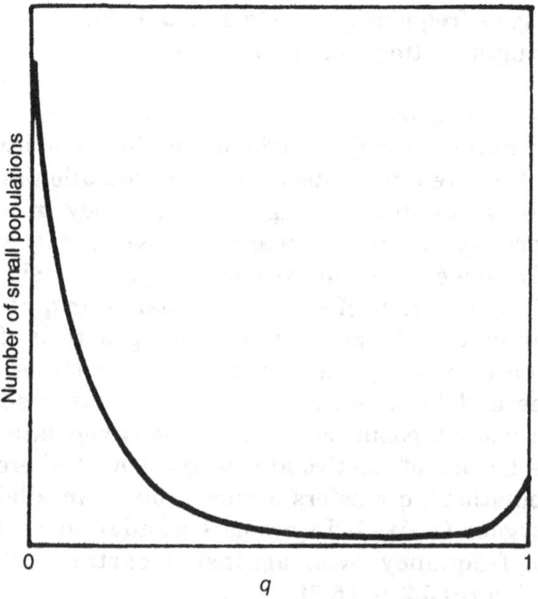

Figure 5.2. Distribution of gene frequency q for gene a in small populations with $4Ns = 5$ (s, selective disadvantage of heterozygotes Aa; $2s$, selective disadvantage of homozygotes aa). In most populations, the gene a will not be present at all. In a few populations, it will attain a moderate or high gene frequency, and in some, it will even replace the other allele (Li, 1955).

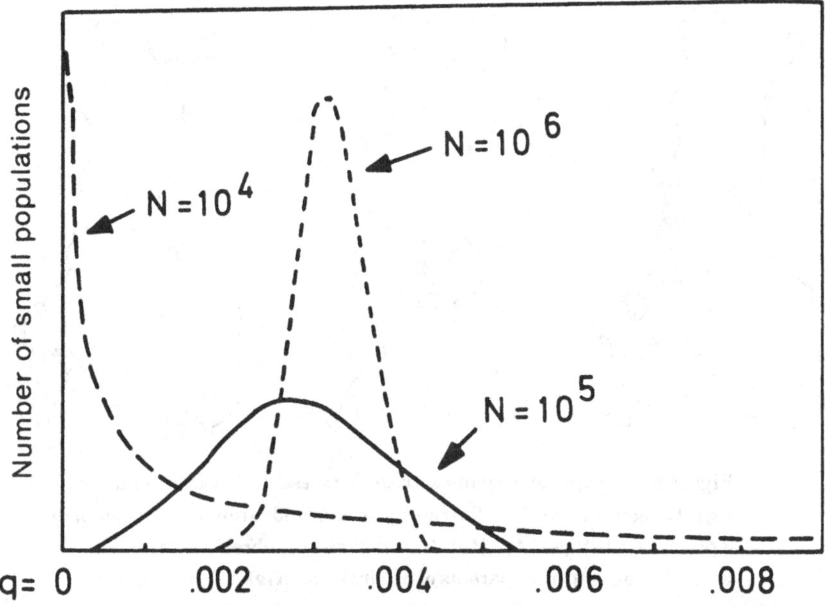

Figure 5.3. Distribution of gene frequency q for gene a in relation to the size N of the effective breeding population with selection $s = 1$ against the homozygotes and mutation rate $\mu = 10^{-5}$. The distribution of gene frequencies q depends critically on the size of the effective breeding population N (Li, 1955).

Classical examples are the various hereditary diseases occurring in the Finnish-speaking population of Finland, such as congenital chloride diarrhoea, nephrotic syndrome, and cornea plana congenita (Figure 5.4). The systematic study of isolates that have persisted up to recent generations, or still persist, has been and is still a powerful tool for the discovery of new recessive diseases. This suggests a research programme in developing countries. The study of persisting isolates for unknown hereditary diseases could provide a genuine contribution to the worldwide development of knowledge in medical genetics - much more useful than mere repetition of complicated laboratory techniques that have been established elsewhere. For example, the People's Republic of China, with its more than 1.1. billion inhabitants, contains within its borders between 50 and 60 different minority groups; many of them are probably still characterised by the old demographic pattern, i.e. living in relatively small, relatively isolated subgroups. Looking for new hereditary diseases is not only

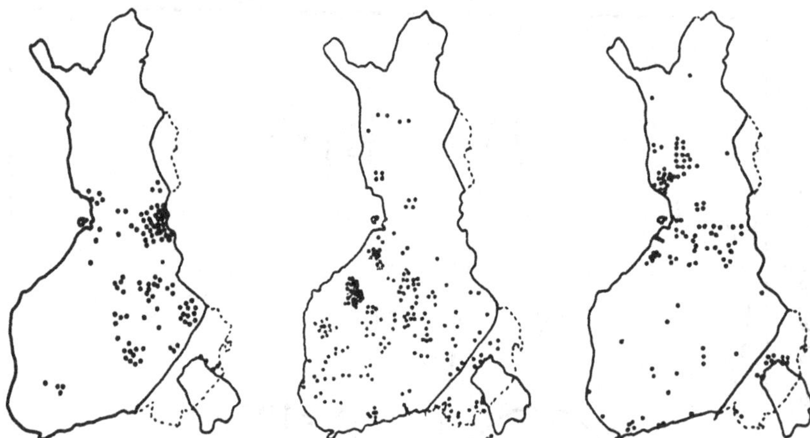

Figure 5.4. Origin of patients with three recessive diseases in Finland. *Left*: Congenital chloride diarrhoea; greatgrandparents of 11 evident and 3 probable sibships (64). *Middle*: Congenital nephrotic syndrome of Finnish type; 60 grandparents of 57 sibships. *Right*: Cornea plana congenita recessiva. Grandparents of 32 sibships. (dashed areas refer to pre-World War II borders) (Norio *et al.*, 1973).

rewarding intellectually for scientists, but also useful from the medical point of view for the populations in which they occur.

Studies of mutant genes at the molecular level - identification of the mutation itself and of haplotypes characterised by surrounding patterns of DNA polymorphisms (RFLPs) - are permitting identification of individual mutational events. Their frequencies and distributions provide useful hints for the reconstruction of population history, and about the existence of isolates in former times. For example, mutants of the phenylalanine hydroxylase (PHA) gene are fairly common in populations of European origin, and many different mutations have been identified (Figure 5.5). The clinical type of phenylketonuria in homozygotes depends, at least in part, on the particular combination of such mutants. In Yemenite Jews, however, only one PHA haplotype has been observed; this population lived for a long time as a relatively small isolate. But isolates have not been, and are still not, limited to peoples living in remote areas, and having little or no contact with the rest of mankind. Even in central Europe, populations were subdivided into subisolates that were separated by boundaries due to geography, religion, social class, and other factors. These

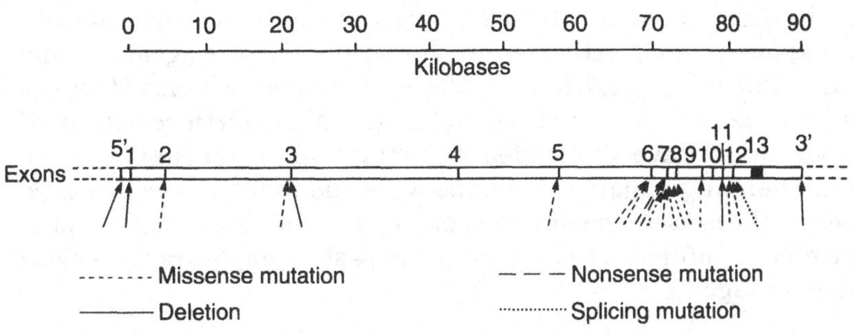

Figure 5.5. The gene for phenylalanine hydroxylase (PHA) and some of its mutations. (Data from Konecki and Lichter-Konecki, 1991).

boundaries are now rapidly disappearing. The most obvious genetic consequence is a levelling-out of frequencies of disease genes. This leads to a decrease in frequencies of homozygotes for such diseases, as a simple calculation shows:

> Assume that in population number 1 homozygotes of the allele a have an incidence of 1:10,000, corresponding to a gene frequency 1:100, and a heterozygote frequency of ca. 1:50, while at another locus homozygotes of allele b occur only once in one million individuals (gene frequency 1:1,000; frequency of heterozygotes 1:500). In population number 2, this relationship is inverted (frequency of allele a 1:1,000, of allele b 1:100). When the two populations mix freely, frequencies will soon become intermediate (ca. 1:180) for each gene, but clinical manifestation reduces to 1:33,000 for the homozygotes for each of them, and 1:16,500 with a gene for both diseases, in comparison with slightly more than 1:10,000 in the two original populations (Vogel, 1986).

Levelling out of gene frequencies leads to a decrease of patients with recessive diseases. Another such decrease is caused by reduction of the proportion of consanguineous marriages, as observed in modern populations. Populations who have gone through the demographic revolution are now experiencing an especially lucky situation as regards recessive diseases: there are good reasons to assume that they are now rarer than ever before.

Consideration of isolate break-up only with regard to recessive diseases is certainly incomplete. Homozygosity also influences traits such as general health, susceptibility to so-called 'multifactorial' diseases, inter-individual genetic diversity of polymorphisms, and possibly even personality characteristics such as intelligence (Vogel & Motulsky, 1986). But these aspects will not be discussed here. Instead, consequences in another field will be explored: interaction of humans with infectious agents. Only about 200 years ago, less than 50% of all newborns reached an age of about 20, i.e. the age of reproduction. While birth defects of all kinds must have demanded a certain toll of casualties, the overwhelming majority of deaths were due to infectious diseases. Hence, the present genetic composition of human populations must have been influenced strongly in the past by exposure to various infective agents.

Man and infection

This conclusion has generally been accepted. However, the reverse conclusion has been rarely considered. The population genetics of infective agents is influenced strongly by genetic and demographic changes of the human host. The interactions of the human host with infective agents form a biological system that must be analysed as a whole.

I shall not offer methods for such an analysis but shall try to characterise the most important elements on which it should be based. First, the 'strategies of defence' on the side of the human host will be compared with the 'attack strategies' of germs. Then, the influence of elements of the demographic revolution on the interaction between host and infective agents will be sketched in crude outline.

Table 5.1. *Immune response*

1.	Binding of antigen by macrophages and presentation to T cells
2.	T cells trigger T-cell cascade
3.4.	Lymphokines induce B-cell proliferation
5.	Immunoglobulins are fixed to antigens and attract complement and killer cells
6.7.	Complement system activates mast cells
8.	Inflammation
9.	Cells loaded with immune complexes are killed by macrophages
10.	In parallel: T cells stimulate killer cells and macrophages
11.	Activation of granulocytes
12.	Infective agents and infected cells are destroyed (killer cells)
13.	Debris is removed (macrophages)

Table 5.2. *C3 phenotype distribution in IgA and Non-IgA-Glomerulonephritis* (Rauterberg, personal communication)

Phenotypes		F		FS		S		χ^2	
		n	%	n	%	n	%		
IgA-GN	observed	13.0	8.0	29.0	17.8	121.0	74.2		
	expected	4.6	2.8	45.7	28.0	112.6	69.1	17.7	$p<0.001$
Non-IgA-GN	observed	4.0	7.4	25.0	46.3	25.0	46.3		
	expected	5.0	9.3	22.9	42.4	26.0	48.1	n.s.	
Controls	observed	14.0	4.4	96.0	30.4	206.0	65.2		
	expected	12.2	3.9	99.7	31.6	204.2	64.6	n.s.	

The immune system of the human host performs its duty of defence against infective agents in several steps, as shown in Table 5.1. Genetic mechanisms providing the functions necessary for this defence system range from the simple to the very complex. Complement factors may be mentioned as examples of a simple mechanism. Some of them show conventional genetic polymorphisms. Influence of polymorphisms on function is shown by their association with diseases (Brönnestam, 1973; Farhud *et al.*, 1972; Table 5.2). The MHC traits, on the other hand, which are necessary for the presentation of the antigen to the T-helper and T-killer cells give a more complex picture. Here, the enormous degree of inter-individual genetic variability at the HLA loci is remarkable. At the protein level, this leads to inter-individual differences in structures for recognition of antigens. The biological significance of these differences is indicated by the well-known associations between HLA types and diseases (see Tiwari & Terasaki, 1985). Here the conventional genetic defence strategy - polymorphism in populations - shows a high degree of refinement. Linkage disequilibrium is the only slightly unconventional element in this component.

On the other hand, the genetic basis for the steps that follow -

presentation of the antigens to T-helper cells and formation of immunoglobulin - is very unconventional. For the T-cell receptors as well as for the immunoglobulins, more generally available cell recognition genes are used after genetic adaptation to this special function (Williams & Barclay, 1988). In addition to constant segments, the respective molecules also consist of variable parts that are products of multiple, very similar but not identical genes, of which only a few and always a different combination are active in a certain cell clone. By this principle, an enormous additional diversity and flexibility of defence mechanism is created. This system calls to mind the elastic defence strategy used in modern warfare which also combines a system of agents in depth, hierarchically ordered defence lines with devices for counter attack. The same military metaphor can also be used for describing the 'strategies of attack' of infective agents summarised in the volume *Immunsystem* (Anonymous, 1988). A great variety of such agents exist - parasitic, multi- and uni-cellular animals, fungi, bacteria and viruses - to mention only a few.

Intestinal worms, for example, are widespread in many populations; under primitive living conditions they contribute appeciably to child mortality. In the blood of worm-infested patients, numerous eosinophilic granulocytes are found, pointing to increased IgE formation. An increase of eosinophilic cells and IgE is also seen in patients with so-called atopic diseases, e.g. endogenous eczema, bronchial asthma and hay fever. In a New Guinea population it was found that the degree of hookworm infection as measured by the number of hookworm eggs in the stool was significantly lower in atopic individuals than in the general population (Grove & Forbes, 1975). This selective advantage might explain the high prevalence of genetic liability to atopies in human populations.

Schistosomiasis is another example. Schistosoma are worms that are propagated in certain water snails. They enter the human host through the skin, leading to a complex disease. To cope with the immune system, they have evolved a double membrane to cover themselves. Damage caused by the human immune response can be repaired fairly simply by abandoning and replacing the damaged parts of the outer membrane. The animal is also able to build antigens like those of the host, such as ABO and MHC structures. The immune system is deceived by these: it wrongly recognises these agents as belonging to the host and therefore does not mount an attack on them.

Further down the biological scale are the trypanosoma (Donelson & Turner, 1988). They cause sleeping sickness in humans and Nagana disease in cattle. Trypanosoma are unicellular organisms somewhat bigger than erythrocytes. They spend long periods of their life cycle in the blood; not within protecting cells as do the malaria parasites and, in part, tuberculosis and leprosy bacteria, but in the free blood stream. They protect themselves from assault by the immune system by continuously exchanging their coantigens. The germ has hundreds and possibly even thousands of genes for its surface glycoproteins, the expression of which varies according to certain rules. While antibody formation in the host is triggered by intrusion of a certain clone of trypanosoma with identical surface antigens, a few individuals have changed expression; their surface glycoproteins differ too much from the majority that they escape antibody attack.

This change of antigens is a widespread principle in various infective agents. In trypanosoma it is made possible by the fact that very many genes are available for one function - genes that are in ever-changing use. Here the same principle is used in attack by infective agents as in defence by the host, for example in immunoglobulin and T-cell receptor determination.

In bacteria again, a number of genetic mechanisms for overcoming host defence are encountered. Interactions mediated by carbohydrates of cell surfaces are relatively simple: some *E. coli* strains that are normally present in the human colon have surface antigens that are so similar to ABO blood group antigens that antibodies will cross-react. This similarity has important consequences in susceptibility to and cause of some diseases - for example *E. coli* enteritis of young children. Various types of enteritis and diarrhoea are still major causes of child mortality in many developing countries. Similar mechanisms have very probably been acting in smallpox as well as in plague and possibly cholera (Vogel & Motulsky, 1986).

Another aspect of bacterial strategies is their ability to code resistance factors, not in the bacterial chromosome itself but in plasmids. Plasmids can be transferred easily to other bacterial cells; in the presence of a selective factor, for example an antibiotic, a bacterial population can become resistant much faster than by the more conventional method of selective advantages of resistance mutants.

Various genetic strategies are also used by viruses. The HIV virus, to mention one example, belongs to the retroviruses which use RNA as genetic material and a template for formation of a double-stranded copy of their genome. This copy is then integrated in the host cell genome. There is also a great deal of genetic variability, but it is brought about by a different mechanism: a certain lack of precision in formation of the DNA copy, i.e. a high degree of mutability (Koch, 1989). Genomes of HIV clones may differ remarkably. This antigen drift is a very efficient but not very original attack strategy. The most dangerous aspect of the HIV infection concerns the type of cells the virus mainly infects. The T-helper cells are its predominant target. As mentioned above, these cells have a central function in the immune response: the foreign antigen is presented to them and they initiate the T-cell cascade, which leads among other things to proliferation of B-cells and antibody formation. To return to our military metaphor, the HIV infection disables important centres of command.

Strategies of attack used by infective agents will now be compared with strategies of defence used by the host. During evolution, both sides in conflict with each other developed complex strategies, that are today being elucidated step by step. Already a number of general principles can be recognised, the most general and most important of which is that of multiplicity and variability. The host has established a defence system that works at many levels, and with quite different elements. They may be fixed or movable, constantly present or made available upon demand. Variability is intra-individual as well as inter-individual. Maintenance of the efficiency of such a system depends on continuing natural selection, since random mutations would gradually destroy it.

The same general principle - creation of multiplicity and variability - is also found on the side of infective agents. Examples are the inter-individual differences in antigen patterns in bacteria and viruses. This variability is complemented by intra-individual variability - antigen drift in trypanosoma may be repeated as an especially well-developed example.

The molecular mechanisms creating this variability are surprisingly similar in host and infective agent, probably mainly because the genetic material can react to a variety of challenges only by a restricted number of answers:

1. germ cell mutations leading to certain protein changes;
2. somatic mutations leading to different cell clones within the same individual or clone;
3. integration of parts of foreign genomes or gene products;
4. formation of multiple copies of the same gene, together with a switch mechanism leading to gene activity of only one or a few of these copies within one cell clone.

These complex systems are of relevance to the breaking-up of isolates. Isolates are characterised by a relatively high degree of homozygosity due to inbreeding and random fixation of genes. This reduces inter-individual variability. Intra-individual variability is also reduced, but to a lesser degree. Even with a high level of homozygosity, intra-individual variation - for example in antibody or T-cell receptor formation - will still be formidable. Keeping in mind that evolution at least of higher primates has occurred in small groups, and that high variability is an advantage for the defence against varying infective agents, this may be one reason for the evolution of mechanisms of increasing variability that were independent of population structure, such as antibody or T-cell receptor determination.

As regards infective diseases, human isolates will harbour a somewhat restricted array of agents. When ecological conditions are constant over many generations, distribution of agents will, as a rule, also remain constant. Mutual adaptation will lead to a kind of equilibrium between both components. The host will continually pay a certain toll of individual morbidity and mortality but the group will survive. The immune defences of the host will impose a strong selection against the infective agent; this in turn will lead to refinements of its genetic attack mechanisms so that the agent also will survive.

But this equilibrium is a precarious one. A certain mutation may turn a relatively benign agent into a very dangerous one, or a slight change of ecological conditions may lead to an unusual increase of a certain infective agent. In this way a human group may be affected severely by disease or may even become extinct, destroying together with itself the infective agent as well. By and large, however, a relatively stable situation will exist.

Effects of break-up of isolates

Break-up of isolates will lead to a number of far reaching changes on the part of the human host as well as of the infective agents. The short-term genetic consequences for the human host when infections are not considered will largely be advantageous. Decrease of homozygosity for recessive genes means decrease of incidence of certain hereditary anomalies and diseases, and increase of inter-individual variability in immunological defence mechanisms. Prediction of long-range genetic consequences is more difficult but some may be disadvantageous, e.g. the number of recessive genes in heterozygotes will increase slowly because not so many of them are eliminated in homozygotes.

Much stronger, however, will be the genetic changes in human populations due to changes in the pattern of infections, and their prediction is much more difficult or even impossible. They depend on the population dynamics of the infective agents. But the following changes can be predicted and have often been observed:

1. Infections that had been endemic before in one or a few small populations will spread into larger populations. This may lead to difficulties in adaptation of newly infected populations of the host, and hence to more severe effects of certain diseases.
2. An extreme consequence is the occurrence of big epidemics. Examples are well known. Plague for example, which is caused by a bacterium, *Yersinia pestis*, has been - and probably still is - endemic in some East African populations, but has often led to worldwide epidemics. The best known was the black death around 1348, which reduced European populations by up to one half. A likely genetic vestige is the higher prevalence of blood group gene O in the more isolated populations of the old world, and its lower frequency in bone material from after the black death in comparison with the time before (Vogel & Motulsky, 1986). Similar examples are the classical epidemics such as smallpox or cholera, but also modern and sometimes less conspicuous ones such as the various waves of *E. coli* infections leading to infant diarrhoea, which were experienced in Europe in the 1950s and early 1960s.

   The latest example of this type is the HIV (see Koch, 1989). To the best of our knowledge, the same or a very similar virus was present in a certain part of Central Africa for a long time. Due to changes in population structure, migration and behaviour, it has spread during recent decades through much of the present world population. As mentioned, this infection is especially efficient in invalidating the defence system of the host, affecting mainly the T-helper cells - a centre of command of the immune system. Effective therapy is not available; therefore, we are now experiencing a similar - though much milder - situation as our ancestors did at the time of the black death.
3. So far, we have not considered the fact that breaking up of isolates is often combined with thorough changes in ecological conditions. In the history of the human species, the neolithic revolution several thousand years ago was the most important event (except for the changes now taking place). Among many other changes, replacement of the hunter-gatherer way of living by agriculture led to a widespread reduction of isolating mechanisms between populations, to an increase in population size, and to partial break-up of isolates. Large parts of the

tropical rain forest were cultivated, and permanent settlements based on agriculture were founded. As analysed by Livingstone (1958, 1967), this led to the formation of open ponds, in which the malaria mosquito found ideal conditions for multiplication. As a consequence, tropical malaria became hyperendemic. The human population adapted itself to a certain degree, and managed to survive by increased frequencies of mutants, leading to an improved resistance of erythrocytes against *Plasmodium falciparum*. Examples are the high incidence of HbS and C in parts of Africa, HbE in southeast Asia, and the various thalassaemia mutants in malarial areas (Vogel & Motulsky, 1986).

4. In modern times, isolates are breaking up completely, inbreeding is being reduced still more, and more effective antibiotics and drugs against the known infections are becoming increasingly available. This has two consequences. On the one hand, natural selection due to the classical infections has decreased steeply. This may lead to accumulation of slightly deleterious mutants within the numerous genes determining the immune system, bringing long-term consequences for resistance not only against infections, but against tumor diseases as well. On the other hand, there will be the danger that the successful fight against the classical infections will open the way for the appearance of new infections with unpredictable consequences.

Conclusion

Human population genetics has developed into a specialised branch of science in which the main forces of evolution, namely selection and random changes of gene frequencies, are usually analysed singly and in their interaction. The same is true in principle for population genetics of other types of life, for example infective agents. But it is often forgotten that human beings and infective agents together form a complex system. It should be analysed together - and with all its interactions. This system is embedded in the still more complex ecosystem. This integrated approach has been illustrated here.

In summary, we started with some aspects of classical population genetics in relation to break-up of isolates. We repeated the well-known results that homozygosity for 'deleterious' mutants will be reduced - a favourable short-term effect for the population. Then we compared genetic defence mechanisms of the human host against infective agents with the genetic mechanisms developed by these agents for attack against the host. As a next step, we considered the break-up of isolates from the viewpoint of infective agents, ranging from a precarious equilibrium to disturbances by worldwide epidemics, and to the artificial situation in a world population in which isolates have disappeared or are disappearing quickly, and in which modern hygiene, antibiotics, drugs and other means of fighting infections are increasingly available.

In this way, some conditions for a new population genetics have been formulated and some of its elements have been named. This population genetics should develop models and analyse actual situations, considering both infective agents and the human host in a unified approach, a true ecogenetics.

References

Anonymous (1988). *Immunsystem*, 2nd edn, Heidelberg: Spektrum der Wissenschaft.

Brönnestam, R. (1973). Studies on the C3 polymorphism: relationship between C3 phenotype and rheumatoid arthritis. *Human Heredity*, **23**, 206–213.

Donelson, J. E. & Turner, M. J. (1988). Wie Trypanosomen das immunsystem täuschen. In *Immunsystem*, 2nd edn, pp. 174–183. Heidelberg: Spektrum der Wissenschaft.

Farhud, D. B., Ananthakrishnan, R. & Walter, H. (1972). Association between the C3 phenotypes and various diseases. *Human Genetics*, **17**, 57–60.

Grove, D. I. & Forbes, I. J. (1975). Increased resistance to helminth infestation in an atopic population. *Medical Journal of Australia*, **1**, 336–338.

Koch, M. G. (1989). *Aids: vom molekül zur pandemie*. Heidelberg: Spektrum der Wissenschaft.

Konecki, D. & Lichter-Konecki, U. (1991). The phenylketonuria locus: current knowledge about alleles and mutations of the phenylalanine hydroxylase gene in various populations. *Human Genetics*, **87**, 377–388.

Li, C. C. (1955). *Population Genetics*, University of Chicago Press.

Livingstone, F. B. (1958). Anthropological implications of sickle cell gene distributions in West Africa. *American Anthropologist*, **60**, 533–562.

Livingstone, F. B. (1967). *Abnormal hemoglobins in human populations*. Chicago: Aldine Press.

Norio, R., Nevanlinna, H. R. & Perheentupa, J. (1973). Hereditary diseases in Finland; rare flora in rare soul. *Annals of Clinical Research*, **5**, 109–141.

Tiwari, J. L. & Terasaki, P. I. (1985). *HLA and disease associations*. New York: Springer-Verlag.

Vogel, F. (1986). Sind Rassenmischungen biologisch schädlich? In *Der ganze Mensch*, ed. H. Rössner, pp. 92–109. München: Deutscher Taschenbuch-Verlag.

Vogel, F. & Motulsky, A. G. (1986). Human genetics: problems and approaches, 2nd edn. Berlin: Springer Verlag.

Williams, A. F. & Barclay, A. N. (1988). The immunoglobulin superfamily – domains for cell surface recognition. *Annual Review of Immunology*, **6**, 381–405.

6 *Isolates in India: their origin and characterisation*

KAILASH C. MALHOTRA

The peopling of India

The discovery of lithic and bone tools of the Early Stone Age and subsequent periods from almost all over the country establishes beyond doubt that man has existed for a very long time in India, and the recent discovery of osseous remains of *Homo erectus* in central India shows the presence there of his antecedents also.

Archaeological and historical evidence further establishes that several waves of immigrants from west and central Asia, and to a lesser extent from north and southeast Asia, came to India during at least the last 10,000 years. The immigrants came primarily by land, but also used sea routes, and mainly settled in the river-valleys and plains. These immigrants brought a variety of cultural as well as biological traits. There is perhaps no other sub-continent in the world that harbours such a rich array of ethnic elements as India. In classical terms all the major races of mankind - Europoid, Mongoloid and Negroid - have contributed in the biological constitution of the people of India (Malhotra, 1978). Numerous anthropometric studies conducted on contemporary Indian populations, and on the prehistoric skeletal remains from different periods and geographical areas, suggest the existence of, or the presence of contributions from, the following ethnic elements among the people of India:

- i Negritos (Andaman Islands)
- ii Australoids (west, central and southern India)
- iii Europoids (all over the country)
- iv Mongoloids (north-east and sub-Himalayan region)
- v Negroes (west coast and southern India, numbering around 15,000)

In terms of antiquity, the evidence both cultural and biological suggests that the Australoids on the mainland, and Negritos in the Andaman Islands are probably relict groups from the oldest inhabitants of the country. Europoids came to India in several waves stretching over the last 10,000-15,000 years. The antiquity of Mongoloids is not yet fully established but perhaps they came in

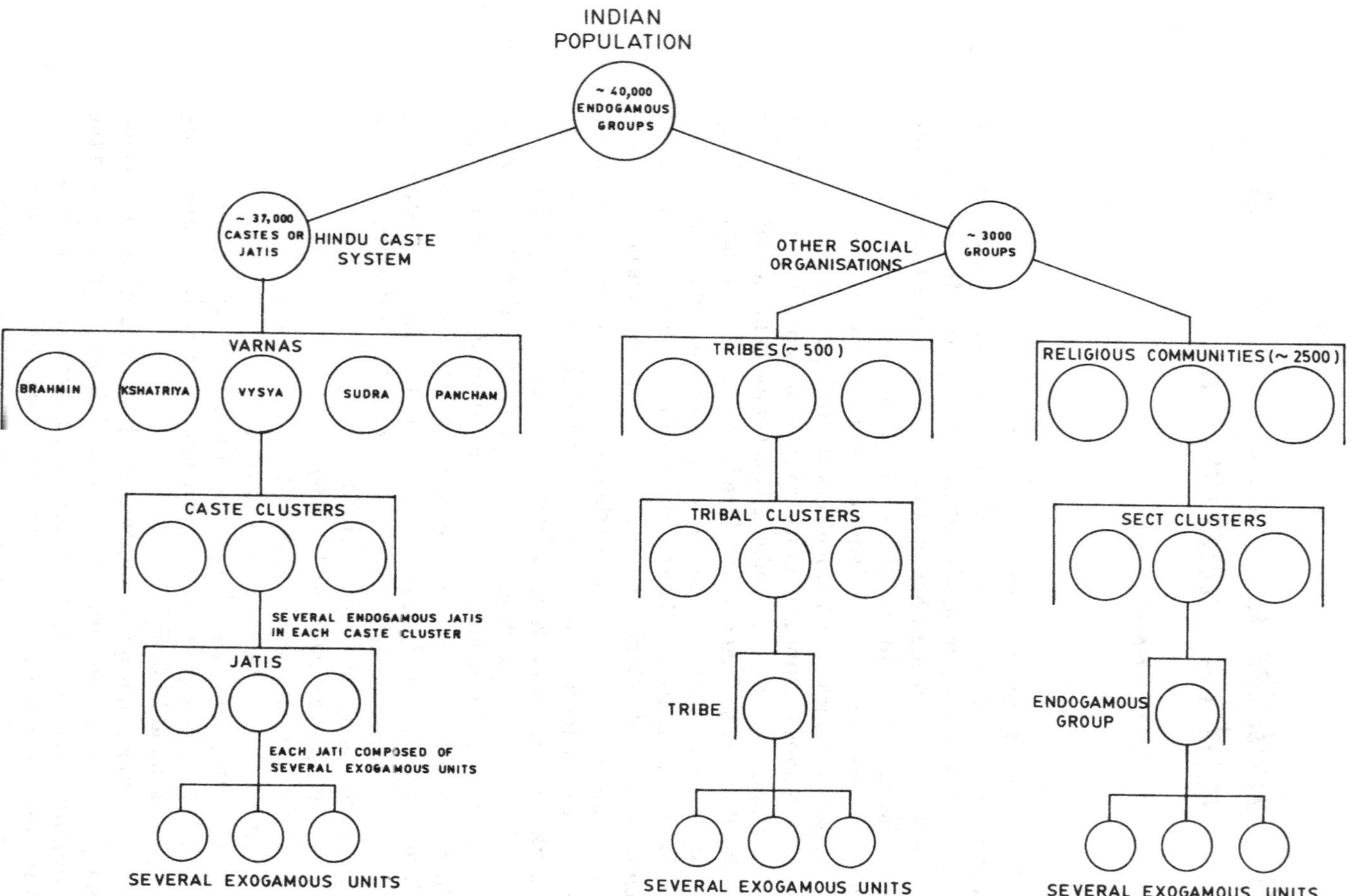

Figure 6.1. Population structure of Indian populations.

recent times, at least in several parts of the north-east. The Negroids from Africa came during the historical period beginning 16th century AD.

It is therefore expected that the considerable amount of variation that exists in morphological characters will also be discerned in genetic traits among the people of India.

Emerging social organisations

The process of peopling of India brought not only different biological elements, but also a wide variety of cultural traits, and the historical and social processes led to the evolution of a unique social organisation, the Hindu caste system in the sub-continent. This social institution has an important bearing on the understanding and interpretation of genetic polymorphisms in the country.

Of over 40,000 Mendelian populations that make up the people of India (Malhotra, 1984), an estimated 37,000 endogamous groups are structured in the Hindu caste system. Each population in the system is called a jati or a caste (Karve & Malhotra, 1968). The 3,000 Mendelian populations that, strictly speaking, are outside the caste system include the tribal populations, and the religious communities like Muslim, Christian, Buddhist, Jain and Sikh. The main features of the Indian population structure are summarised in Figure 6.1.

Each caste theoretically belongs to one of the five varnas which are arranged in a hierarchical order. The Brahmin varna is at the top of the social hierarchy, followed by the Kshatriya, Vaishya, Sudra and Pancham varnas. Traditionally, the members of each caste followed a particular profession, practised endogamy, and had elaborate rules for selecting a mate.

Consequences of the social structure

Some of the more notable consequences of the Hindu caste system, in terms of micro-evolutionary implications, have been:

a. the entire population has been divided into a large number of endogamous groups and the system did not permit large-scale intercaste marriages, so that the gene pool of each caste has evolved separately over the last 3,000 years;
b. the population sizes of the castes show enormous variation; there are several with fewer than 10,000 persons, whereas many others number over 100,000 people;
c. several castes are located in rather small territories, often only in a single village, whereas many others are found in several districts;

d. there is uneven geographical distribution of many castes because of occupational differences (Malhotra, 1980);
e. there was retention of diversity in many sociocultural practices including those related to the choice of mates among different castes.

This social organisation poses certain methodological issues related to the sampling procedures and units of study (Malhotra, 1974). It suffices here to say that in the Indian situation it is necessary that the unit of study should be the endogamous group in view of the fact that castes have evolved their own gene pools separately. Plenty of evidence exists to show that several genetic traits do not occur randomly in a region, but often are confined to specific endogamous groups. Further, since several castes are very small numerically, models used for analysing large populations cannot be applied for such populations. Such a situation occurs even in a small geographical area. There is a bewildering array of patterns in the choice of mates, from the practice of various types of consanguineous marriage to avoidance of consanguineous marriages. Briefly, the Indian social structure, compared to structures in many parts of the world, poses unique and special methodological issues.

Genetic polymorphism in Indian populations

Although studies on blood groups started on Indian populations in the mid-twenties, systematic studies began only after the publication of a paper by Sanghvi and Khanolkar in 1949. Since then a large number of populations - tribes, castes, religious isolates - from almost all parts of the country have been studied. Although periodic reviews have been attempted in the past (e.g. Bhalla, 1984), systematic compilation and synthesis of the large amount of data that are now available has yet to be attempted. Two major recent efforts in this direction have been the *Genetic Atlas of Indian Tribes* by Bhatia and Rao (1986) and the monograph *The Distribution of Human Genetic Polymorphism in India* by Cassero *et al.* (1986). In the absence of a comprehensive compilation, it is, therefore, not possible to give here details of the amount of work that has been done in India, but a few examples will illustrate the extent of the data that have been accumulated over the years.

Of the approximately 500 tribes in the country, 191 tribes have been studied for at least one of the following genetic systems. Many non-tribal populations have also been studied for these:

a. Blood groups — A_1A_2BO, MN, Rh, In(a), Duffy, Kell, Lutheran, Kidd, P, Lewis, Diego, ABH secretion. Over 900 populations had been studied for ABO blood groups up to 1980.

b. Red cell enzymes — Esterase D, acid phosphatase, adenylate kinase, Adenosine deaminase, phosphoglucomutase I and II, lactic dehydrogenase, malate dehydrogenase, 6-phosphoglucomutase dehydrogenase, glyoxalase I, peptase A, B and C.

c. Serum proteins — Albumin, haptoglobin, transferrin, group specific component, pseudocholinesterase E_2, lipoprotein, Gm, Inv. groups. Mukherjee and Das (1984) reported that 148 populations had been studied for haptoglobins up to 1978.

d. Other systems — G-6-PD deficiency, haemoglobinopathies, PTC, colour-blindness. Over 150 populations have been studied for haemoglobinopathies.

Until 1960 most of the work pertained to serological systems (in particular ABO, MN, Rh), PTC, colour-blindness, ABH secretion, G6PD and haemoglobinopathies, then studies on enzymes and proteins increased during the seventies. In the eighties, studies on the HLA system and increased numbers of enzymes and proteins were initiated, including the use of the isoelectric focussing technique. Recently a few studies have been published using restriction fragment length polymorphisms (RFLPs). A number of well planned studies at regional level are now available - Assam, Bengal, Maharashtra, western India, Tamil Nadu, etc. in which several groups representing various strata of the society were studied for a large number of genetic systems.

For a number of polymorphisms in the Indian population, in terms of geographical, social and ethnic divisions, the range of variation in gene frequencies of the tribals and Hindus (undifferentiated by social hierarchy) are given (Table 6.1). The main feature that emerges is that there is a wide range of gene frequency in all genetic systems, and in general the tribes and Hindu castes show a great deal of variation, specially for the alleles P_1, D, C, E, Se, Fya, Gd, $PGM_{2\text{-}1}$, AK, EstD, PTC, etc.

Ethnic variation

Several studies are available of genetic relationships between the tribes and non-tribal populations based on a large number of loci, e.g. Balakrishnan (1984) using data on ABO and Rh subtypes

Table 6.1. *Range of variation in gene frequencies for selected genetic systems among the Hindu and the tribal populations of India*

Genetic system	*Gene symbol*	*Hindus*			*Tribals*		
		min	*max*	*mean*	*min*	*max*	*mean*
ABO blood groups	A	0.075	0.302	0.190	0.142	0.355	0.193
	B	0.118	0.313	0.245	0.118	0.340	0.256
	O	0.421	0.708	0.570	0.412	0.657	0.552
MN blood group	n	0.324	0.475	0.378	0.229	0.422	0.349
P blood group	p_1	0.221	0.767	0.459	0.045	0.587	0.368
Rh blood groups	D	0.684	0.944	0.774	0.743	0.980	0.825
	C	0.605	0.810	0.679	0.542	0.856	0.741
	E	0.067	0.142	0.103	0.069	0.220	0.116
Duffy blood group	Fy^a	0.259	0.684	0.441	0.414	0.735	0.557
ABH Secretor	Se	0.390	0.549	0.480	0.445	0.641	0.544
Hp	Hp^1	0.110	0.217	0.160	0.103	0.331	0.175
Acid phosphatase	P^a	0.188	0.259	0.239	0.117	0.415	0.231
G-6-PD	Gd-	0.022	0.046	0.033	0.013	0.140	0.083
Phosphoglucomutase	PGM^1	0.189	0.349	0.289	0.296	0.383	0.327
Adenylate kinase	AK^2	0.089	0.123	0.101	0.041	0.064	0.053
Esterase D	EsD2	0.155	0.453	0.303	0.234	0.385	0.281
PTC	T	0.497	0.810	0.602	0.523	0.724	0.657

among 102 populations; Roychoudhury (1984) in western, eastern, central and south India; Mukherjee *et al.* (1979) in Maharashtra.

The main conclusions that emerge from these studies are:

a. The tribal and non-tribal populations cluster separately, even when gene frequency data of five polymorphic loci are used. This indicates that the tribes in general are genetically different from the non-tribal, though many of them have been living in the neighbourhood of non-tribal populations for a long time.

b. The genetic distances among tribals and non-tribals are correlated more with geographic proximity than with linguistic affinity; the tribes of Bihar, Orissa and eastern Madhya Pradesh are genetically closer to one another than to those of Gujarat and Madhya Pradesh or to those of the Nilghiri Hills of south India.
c. The Mongoloid tribes in northeastern India are in general characterised by the absence of A_2 and Rh (d) genes. The tribes of central and southern India possess the HbS gene.
d. The Australoid tribes appear to have been the first settlers in the country.

Social variation

The question of whether the four Hindu varnas are based on genetic similarity of their constituents or whether they are social categories that have evolved through historical and social processes has been attacked in numerous studies that have been carried out in the country. They show:

a. Often castes belonging to one varna are genetically closer to castes of another varna;
b. Castes, irrespective of social hierarchy, often show closer affinities based on region rather than social ranking;
c. The component of total genetic diversity among castes is usually of the order of 1 to 3%, whereas between castes and tribes it is about 5%.

These results show that the variation observed among castes belonging to the different varnas is indeed quite small, and that varnas are not based on genetic similarity between their constituent castes. This confirms earlier observations based on morphology that castes essentially are populations of Europoid origin. That they were derived mostly from west Asia is supported by their close genetic affinities with the Iranians and Afghans.

References

Balakrishnan, V. (1984). Admixture as an evolutionary force in populations of the Indian sub-continent. In *Human Genetics*, ed. K. C. Malhotra & A. Basu, pp. 103–145. Calcutta: Indian Statistical Institute.

Bhalla, V. (1984). Gene diversity in tribal populations of India: Illustrative maps showing its distribution of ABO, MNSS and Rh polymorphism. In *Human Genetics*, ed. K. C. Malhotra & A. Basu, pp. 420–448. Calcutta: Indian Statistical Institute.

Bhatia, H. M. & Rao V. R. (1986). *Genetic atlas of Indian tribes*. Bombay: Institute of Immunohaematology.

Cassero, C., Montana, R., Arena, M. & Modiano, G. (1986). *The distribution of human genetic polymorphism in India*. Roma: Accademia Nazionale Dei Lincei.

Karve, I. & Malhotra K. C. (1968). A biological comparison of 8 endogamous groups of the same rank. *Current Anthropology*, **47**, 109–124.

Malhotra, K. C. (1974). Some models for the study of human population genetics in India: a review. In *Human population genetics in India*, ed. L. D. Sanghvi, V. Balakrishnan, H. M. Bhatia, P. K. Sukumaran & J. V. Undevia. Bombay: Indian Society of Human Genetics.

Malhotra, K. C. (1978). Morphological composition of the people of India. *Journal of Human Evolution*, **7**, 45–53.

Malhotra, K. C. (1980). Gene dispersion in man: the Indian case. *Current Anthropology*, **21**, 135–136.

Malhotra, K. C. (1984). Population structure among the Dhangar caste-cluster of Maharashtra India. In *The people of South Asia*, ed. J. R. Lukacs, pp. 295–323. New York: Plenum Press.

Mukherjee, B. N. & Das S. R. (1984). Haptoglobins: genetics and variation in Indian populations. *Indian Journal of Physical Anthropology and Human Genetics*, **10**, 96–120.

Mukherjee, B. N., Majumder, P. P., Malhotra, K. C., Das, S. K., Kate, S. L. & Chakraborty, R. (1979). Genetic distance analysis among nine endogamous population groups of Maharashtra, India. *Journal of Human Evolution*, **8**, 567–570.

Roychoudhury, A. K. (1984). Genetic relationship of Indian populations. In *Human genetics*, ed. K. C. Malhotra & A. Basu, pp. 146–174. Calcutta: Indian Statistical Institute.

Sanghvi, L. D. & Khanolkar V. R. (1949). Data relating to seven genetical characters in six endogamous groups in Bombay. *Annals of Eugenics*, **15**, 52–75.

7 *Consanguineous marriages and their genetical consequences in some Indian populations*

D. P. MUKHERJEE

Introduction

With the microevolutionary theory emerged the concept of a population as an isolated group of conspecific interbreeding individuals who share a gene pool. But human populations present a hierarchy of further isolations. Consanguinity in marriage, where it occurs, serves as an isolating mechanism at the lowest level. Such marriages have important implications for the health and wellbeing of the subsequent generation, and recognition of this led to systematic studies on the rates and patterns of marital consanguinity in different countries. The highest rates so far have been found in Japan, India, Israel and Brazil. But there is wide variability within each country and within local populations. The variability and the trends of consanguineous marriage in Indian populations are mainly determined by the marriage regulations and cultural traditions.

Marriage regulations

The first level of breeding isolation in Indian populations is determined by religious and sect affiliations and linguistic-cultural diversities and traditions. About 75% of the people make up the traditional Hindu society with its network of varnas and endogamous jatis, about 8% belong to more isolated and relatively tradition-bound groups called the scheduled tribes near forests and hills, about 13% belong to Muslim communities, and the rest to Sikh, Christian, Buddhist, Jain, Parsee and Jewish. All of them are distributed over the different provinces and states, with different linguistic or other traditions. These populations originated mainly from indigenous peoples (Hindus) and are divided and subdivided into endogamous occupational, local and cultural units (Mukherjee, 1971). In each endogamous group, specially Hindus, the rule of lineage-exogamy prevents the population from splitting up and prohibits marriage between

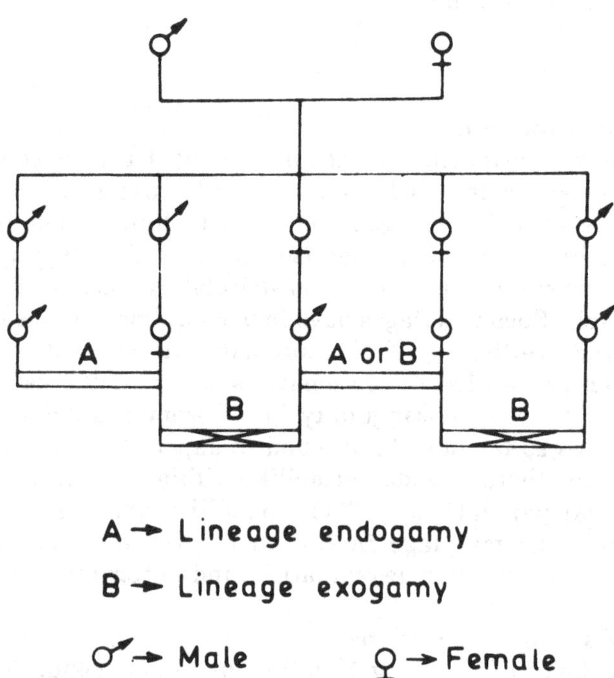

Figure 7.1. Pedigree showing cross-cousin and parallel-cousin marriage.

parallel cousins, offspring of brothers or of sisters (Figure 7.1). In matrilineal populations such as the Khasi, marriage is also restricted between cross-cousins, the offspring of brothers and sisters (Dasgupta, 1964). In many endogamous populations, there is a regulation against consanguineous marriages and specifically between descendants of paternal and maternal ancestors within seven and five generations respectively. The Brahmins in north India influenced neighbouring Hindus directly and others indirectly to adopt this custom, but in southern states this rule was relaxed to accommodate the prevailing custom of close consanguineous marriages (Boudhayana cited in Sanghvi, 1966; Mukherjee, 1971). Southern peoples also practise village-endogamy, in contrast to most parts of the north where village-

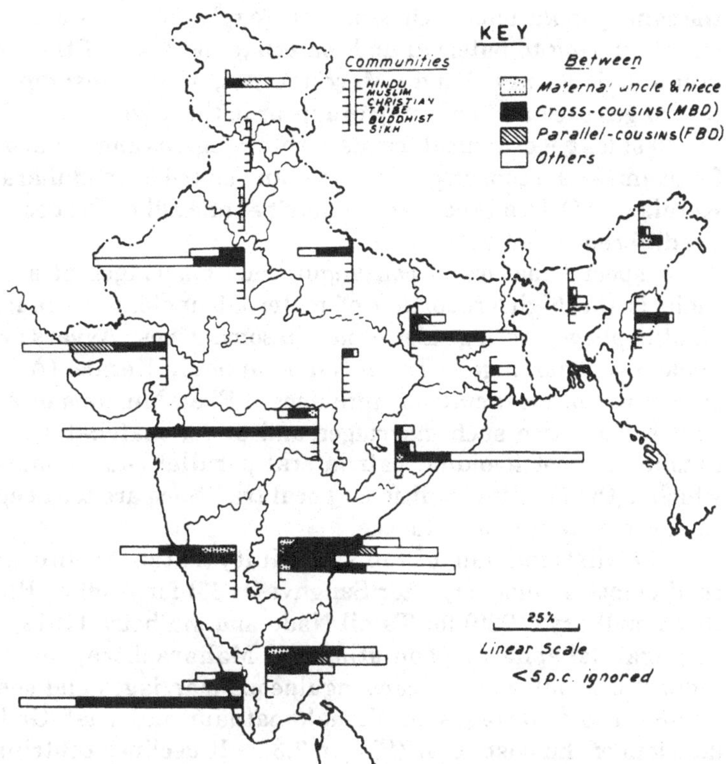

Figure 7.2. Incidence of consanguineous marriages in India by State (based on 1961 census).

exogamy is the rule. In Uttar Pradesh many prefer to bring brides from the east to further limit the chances of consanguinity.

Geographical trends, clines and local influence

Geographical plotting of the data from the 1961 census of India, analysed by Roychoudhury (1976) for broad religious/social categories (Figure 7.2), indicates high consanguinity rates in southern states (except Kerala) and particularly in Andhra Pradesh. The incidence of such marriages declines through neighbouring states, and is greater in the western regions than in the east (as far as the hills of the north-east). It is absent in the sample of 4910 marriages from West Bengal collected with the 1961 census. This reflects the ancient observation (of about 1,500 to 2,000 years ago) that consanguineous marriages are most rigidly avoided in eastern India. It has not been found in local plains tribes of Bengal and Bihar, though sporadic cases occur in the same populations in Orissa. Due to relaxation of the custom of ritual offering to paternal and maternal ancestors of the Sapinda group and the new Hindu Marriage Act, 1955, consanguineous marriages are now beginning to appear in this region.

Besides geographical trends within religious and social groups, for example the consanguinity rates among tribes in Maharashtra, Kerala and Orissa exceed those in tribes of Andhra Pradesh, there are differences between them.

A special feature of consanguineous marriages of southern India is the high proportion of maternal uncle:niece marriage. Aunt:nephew marriage also occurs in some tribes; Ayyars practise uncle:niece marriage in Tamil Nadu but not in Kerala (Ali, 1968) on account of local cultural influence. Even Muslims of Andhra Pradesh practise such marriages and prefer matrilateral cross-cousins, almost avoiding patrilateral parallel cousin marriage, which is the Muslim tradition. Local traditions are too deep to be removed by religious influence.

The distribution of consanguinity rates or inbreeding coefficients by district, after Sanghvi (1966) for Andhra Pradesh, Roychoudhury (1980) for Tamil Nadu and Malhotra (1979) for 22 Dhangar (shepherd) populations of Maharashtra, shows the highest concentration of consanguineous marriages and specially uncle:niece marriages in Vishakapatnam and east Godavari districts of the east coast (Figure 7.3). It declines centrifugally. The high rate of consanguinity along the east coast is interrupted by a middle region of low rates, declining westward from Krishna district. The decline is towards the south and west in Tamil Nadu. Dhangars moving northwards from the south and Karnataka state have given up the practice of uncle:niece and patrilateral cross-cousin marriages while strongly retaining the tradition of matrilineal cross-cousin marriage. The clines in consanguinity rates are also reflected in the data collected from hospital patients of four different localities (Dronamraju & Meera Khan, 1960; Mukherjee, 1972; Murty & Jamil, 1972; Veerraju, 1973).

Variations among Muslims

The highest rates of consanguinity among Muslims are found in Kashmir (Kashyap, 1978), Uttar Pradesh, Rajasthan and Gujarat (Basu, 1975; Roychoudhury, 1980), Andhra Pradesh (Sanghvi, 1966; Mukherjee *et al.*, 1974) and the Muslims among the Negroid Siddi population of Karnataka (Vijaykumar & Malhotra, 1981). The Shia Muslims of the north-west show the highest frequencies (Basu, 1975; Kashyap, 1980) of consanguineous marriages and especially those between offspring of brothers. The incidence of

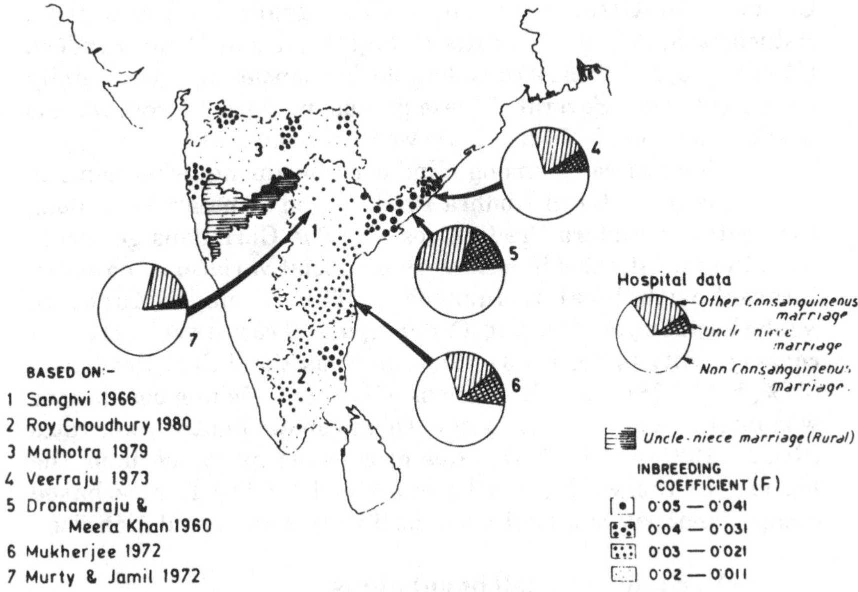

Figure 7.3. Regional clines of consanguineous marriages and F-values in southern India.

consanguineous marriages among Muslim populations also declines eastwards (Huq, 1976; Mukherjee & Chakravarty, 1977).

Social variations

The patterns of variation in consanguinity rates indicate that both cultural traditions and cultural environment play important roles in determining the type and incidence of such marriages. The consanguinity rates also differ in traditional occupational groups and their subpopulations who also conform to the overall geographical trends. The Brahmans and some other castes such as the 'trading castes' and Kshatriyas in Andhra, Kayasthas in Maharashtra (e.g. Dronamraju & Meera Khan, 1963a,b; Sanghvi, 1966; Goswami, 1970; Mukherjee *et al.*, 1974; Srinivasan & Mukherjee, 1976) have the lowest rates in each state or area within a state.

The belief that consanguineous marriages do not occur among Hindus in northern India is belied by positive findings from a number of Hindu or Hindu-Sikh populations. The Jats of Jammu and Kashmir (Kashyap, 1978), Gujjars throughout the cis-Himalayan regions (Bandopadhyay *et al.*, 1986), some Punjabi castes such as Bhatias, Aroras, Khattris and even Brahmans (Bhalla & Bhatia, 1974), some scheduled castes like Jatavs and

Chamars of Uttar Pradesh, a Sikh group in Delhi from Baluchistan, and the Mewatis of Rajasthan and Uttar Pradesh (Mukherjee, 1972) show considerable frequencies of consanguinity among couples. Several of these groups immigrated from what is now Pakistan in 1947 when India was divided.

The highest rates among Hindus occur among fishermen and shepherds in northern Andhra Pradesh and some artisans along the southern Andhra Pradesh coast. The Christians generally have the lowest rates in each region. Social variations, however, interact with local influences. The farmers (Kapu) in Vishakapatnam district (Veerraju, 1973) have rates of consanguinity (49%, F=0.038) as high as the Jalari fishermen (47%, F=0.038). But the sections of the same fishing populations who migrated to the Puri coast of Orissa show slightly lower rates (Reddy, 1981). The high rates of consanguinity, perhaps the highest in the world, can be explained by the family based occupational pattern, semi-nomadic life and small population size.

Variation in small populations

A small artisan population of Padmasale weavers in Tirupathi area of southern Andra Pradesh (Mukherjee, 1984a) shows a consanguinity rate of 70% (F=0.046). But in a weaver population, the Pattusale, which consists of 1,152 individuals, in 256 marriages 49% are consanguineous (Rao & Mukherjee, 1975). But they have a moderate F (0.030), due to a small 1.2 percentage of uncle:niece and grandfather:grandaughter marriages. A third small population (Kummari potters) in Chittoor district (Mukherjee *et al.*, 1974) also has a similar inbreeding intensity (F=0.031) but a lower rate of consanguinity (36%). Each family tries to monopolise its skill for a specific type of pot-making but, as they have to disperse for the same skills, they display an unusual phenomenon of a higher average marital distance for consanguineous couples (17.2km for 64 marriages) than for non-consanguineous (13.2km for 54 marriages).

Only 659 individuals formed the Toto population of West Bengal in the foothills of Darjeeling. All 207 married couples among them were consanguineous (Debnath, 1978) so that, assuming about equal fertility, the mean F was estimated from the pedigrees to be 0.065.

Rural-urban and temporal trends

A few studies suggest lower consanguinity rates in urban areas, but no data from a specific population are yet available. The effect of literacy is uncertain. There are higher consanguinity rates among both rich potato cultivators at one place and poor land labourers in another among Muslims of Burdwan district, West Bengal.

Sanghvi *et al.* (1956) observed a significant decline of consanguinity rate among the Parsees, but not in other populations of Bombay. The results of other studies are far from uniform. An increase of the rate was reported among the Ahmediahs (Kashyap, 1980) and other Muslims of Uttar Pradesh and Delhi (Basu & Roy, 1972). Bhalla and Bhatia (1974) also reported an increasing trend in a Bhatia group. The same is the case of the Mala of Andhra Pradesh (Mukherjee, 1984a). In contrast, Sundar Rao *et al.* (1972) observed a decline in both rural and urban populations of northern Tamil Nadu.

The strong cultural compulsion for marrying matrilateral cross-cousins among the Kuki tribal populations of Manipur, is reported now to be somewhat relaxed (Das, 1976). A trend is also detected to decline in younger generations among four Brahmin populations of Tamil Nadu (Srinivasan & Mukherjee, 1976), among Bengalee Muslims of West Bengal, and Assam (Mukherjee & Chakravarty, 1977). But there is no corresponding decline of inbreeding level in them. With the reduction of consanguineous marriages there is an increase in the number of uncle:niece marriages in the four Tamil Brahmin groups, and reduction of marriages between relatives more distant than first cousins among Muslims of Burdwan.

Genetic consequences

Population size

One important evolutionary effect of consanguineous marriages is the reduction of the breeding size and marriage area. Pedigrees show the splitting up of a single family into several intrabreeding populations through three or four generations. The concentration of consanguineous marriages can be shown in a Muslim pedigree collected by A. Mukherjee (1985) (Figure 7.4). The phenomenon can also be observed through inter-village differences in small areas within endogamous populations (Veerraju, 1973, 1978; Huq, 1976). The average size of a village ranges from 500 to 1,000 individuals distributed in 100 to 200 families. In Burdwan, for example, consanguineous marriages are found concentrated in two groups of villages separated by a number of villages with negligible rates.

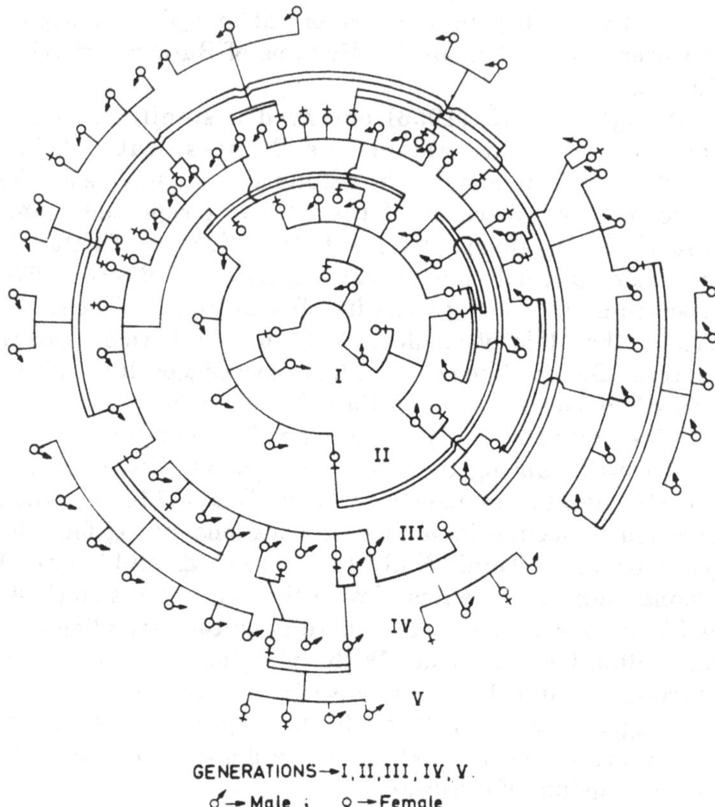

Figure 7.4. Concentration of consanguineous marriages in a Muslim pedigree.

Fertility, mortality and morbidity

In the accumulated data from various endogamous populations in which adequate precautions for minimising environmental and genetical diversities between consanguineous and non-consanguineous couples were taken, and when studies were conducted by the authors themselves (Mukherjee *et al.*, 1974; Reddy, 1983; and others), there is consistent evidence for an enhancement of gross fertility in the consanguineous couples. This agrees with observations in large-scale studies on different groups in Japan (Schull *et al.*, 1968; Schull & Neel, 1972) or among Samaritans (Roberts & Bonné, 1973). But it is only in surveys conducted in Tamil Nadu, which fail to consider breeding structures of the concerned populations, that the results are uncertain in this respect (Sundar Rao & Inbaraj, 1977). The suggestion about consanguinity effects on increased sterility in some Andhra data (Dronamraju & Meera Khan, 1963b) does not hold good in other data (Sundar Rao & Inbaraj, 1977).

There is also a consistent indication of increased incidence of stillbirths and infant mortality in the offspring of consanguineous couples (Mukherjee *et al.*, 1974; Reddy, 1983), but not of abortions, as in the Samaritans (Roberts & Bonné, 1973).

The morbidity rates in terms of congenital malformations appear to increase in consanguineous couples in most populations studied, although rates of increase may not always be very high. It is high in a Kerala study (Kumar *et al.*, 1967) where additional factors may be suggested such as increased natural irradiation from monazite sands which might increase mutation rates, or such as extreme poverty in both consanguineous and non-consanguineous groups which, as Pai (1972) suggests, might enhance expression of the homozygous genotypes of harmful genes.

Sanghvi (1966) formulated a theory of selective elimination of deleterious genes as a result of inbreeding through hundreds of generations. Later analysis, however, revealed 1.49% morbidity in the offspring of consanguineous couples in place of 1.39% in that of the non-consanguineous (Sanghvi & Master, 1974). Dronamraju & Meera Khan (1960) and Mukherjee (1972) observed higher rates of parental consanguinity of patients suffering from various diseases in hospitals of Vishakapatnam and Tirupati respectively, the highest rates being observed in chest diseases and pulmonary tuberculosis. But the results may be compounded with socioeconomic factors. It is more useful to compare specific anomalies. The rates of anencephaly in the stillborn in Tirupati hospital were found to be higher than in Calcutta hospitals (Guha & Mukherjee, 1980). Several Muslim populations of Maharashtra show moderate frequencies of consanguineous marriage (Sanghvi *et al.*, 1956; Malhotra *et al.*, 1977) but among them, as among other Muslims of western India, a large proportion of such marriages occur between patrilateral parallel cousins.

Effects on non-segregating traits

The many consanguineous marriages in Andhra Pradesh Hindus and West Bengal Muslims allow examination of the effects of such marriages on the genetical variation in quantitative physical measurements. Care was taken to select matched controls from same kindreds (Lakshmanudu, 1980; Mukherjee, 1984b) and to select closely related first and second degree relatives who live and work together in the same agricultural fields among Burdwan Muslims (A.Mukherjee, 1985, 1990), so that the observed reduction of mean values in the offspring of consanguineous marriages cannot be ignored on the grounds of genetical and environmental diversities. Consanguinity effects on dermatoglyphic traits like pattern intensity and ridge-count on fingers, which are not at all influenced by environmental factors or socioeconomic status, were also found (Sabarni & Mukherjee, 1975; Mukherjee *et al.*, 1980; Mukherjee, 1984b). The changes of means were proportional to the mean values of non-inbred controls, which strengthens their attribution to a genetical basis. One interesting result of the comparison of the offspring of different degrees of consanguineous marriages is a consistent non-linear change in the mean values, also reported by Barrai *et al.* (1964), and the variances, which are presumably explained by enhanced selection. There was also a trend to bimodality in the distribution of measurements, which are known to be unimodal, among the offspring of related couples, but not so in the offspring of unrelated couples. An antimode also appears to mark the heterozygotes for fingerprint patterns in the offspring of consanguineous parents. These suggest segregation at a few loci and the additive and non-additive effects of the genes concerned. Surprisingly, the consanguinity effects on fingerprint patterns is more marked in the female samples so far studied and this has led to a search for influence of both sex-linked and autosomal genes on the trait. Such consistent consanguinity effects on quantitative traits urge caution in comparing population means and frequencies of appropriate traits without considering the structure of those populations.

References

Ali, S. G. M. (1968). Inbreeding and exogamy in Kerala (India). *Acta Genetica*, **18**, 369–379.

Bandopadhyay, S. S., Chakravarty, S. K. & Lakshmi, G. R. (1986). Consanguinity among pastoral nomad Gujjars of north western Uttar Pradesh. *Human Science*, **35**, 134–140.

Barrai, I., Cavalli-Sforza, L. L. & Mainardi, M. (1964). Testing a model of dominant inheritance for metric traits in man. *Heredity*, **19**, 651–668.

Basu, S. K. (1975). Effects of consanguinity among North Indian Muslims. *Journal of Population Research*. **2**, 57–68.

Basu, S. K. & Roy, S. (1972), Change in the frequency of consanguineous marriages among Delhi Muslims after partition. *Eastern Anthropologist*, **25**, 21–28.

Bhalla, V. & Bhatia, K. (1974). Analysis of migration effect on the mating system of a north India inbred Hindu community – the Bhatias of Garhi – Dhanata. In *Human population genetics in India*, ed. L. D. Sanghvi, V. Balakrishnan, H. M. Bhatia, P. K. Sukumaran & J. V. Undevia, pp. 147–156. New Delhi: Orient Longman Ltd.

Das R. K. (1976). Mother's brother's daughter marriage among the Thadou of Manipur. *Journal of the Indian Anthropological Society*, **11**, 159–168.

Dasgupta, P. K. (1964). Marriage among the War Khasi. *Man in India*, **44**, 146–160.

Debnath, S. K. (1978). Inbreeding among the Totos of West Bengal: a preliminary report. *V. Annual Conference of the Indian Society of Human Genetics, Bombay* (Abstract), pp. 19–20.

Dronamraju, K. R. & Meera Khan, P. (1960). Inbreeding in Andhra Pradesh. *Journal of Heredity*, **51**, 239–242.

Dronamraju, K. R. & Meera Khan, P. (1963a). Study of Andhra marriages: Consanguinity, caste, illiteracy and bridal age. *Acta Genetica*, **13**, 21–29.

Dronamraju, K. R. & Meera Khan, P. (1963b). The frequency and effects of consanguineous marriages in Andhra Pradesh. *Journal of Genetics*, **58**, 387–401.

Goswami, H. K. (1970). Frequency of consanguineous marriages in Madhya Pradesh. *Acta Genetica Medical Gemellol*, **19**, 486–490.

Guha, A. & Mukherjee, D. P. (1980). Trends of congenital malformations in neonates in two Calcutta hospitals with special reference to anencephaly. *Journal of the Indian Anthropological Society*, **15**, 157–166.

Huq, F. (1976). Consanguinity and inbreeding among the Muslims of Murshidabad and Birbhum districts of West Bengal. *Journal of the Indian Anthropological Society*, **11**, 21–25.

Kashyap, L. K. (1978). A new perspective in casual approach to inbreeding: data from five inbred populations of Jammu and Kashmir state. Abstracts. *Xth International Congress of Anthropological Ethnological Science, New Delhi*, p. 39.

Kashyap, L. K. (1980). Trends of isonymy and inbreeding among the Ahmadiyyas of Kashmir. *Journal of Biosocial Science*, **12**, 219–225.

Kumar, S., Pai, R. A. & Swaminathan, M. S. (1967). Consanguineous marriages and the genetic load due to lethal genes in Kerala. *Annals of Human Genetics*, **31**, 141–145.

Lakshmanudu, M. (1980). Inbreeding effects on physical measurements in some Indian populations. Ph. D. thesis, University of Calcutta.

Malhotra, K. C. (1979). Inbreeding among Dhangar castes of Maharashtra, India. *Journal of Biosocial Science*, **11**, 398–409.

Malhotra, K. C., Ahamadi, K., Kazi, R. B. & Bhosa'le, N. (1977). Consanguineous marriages among five Muslim isolates of Maharashtra. *Journal of the Indian Anthropological Society*, **12**, 207–212.

Mukherjee, A. (1985). Inbreeding effects on quantitative traits in a Muslim population of West Bengal. M. Sc. dissertation, University of Calcutta.

Mukherjee, A. (1990). Inbreeding effects on bilateral asymmetry of dermatoglyphic patterns. *American Journal of Physical Anthropology*, **81**, 77–89.

Mukherjee, D. P. (1971). Patterns of marriage and family formation in rural India and genetic implications of family planning. WHO sc. group, pp. 1–18. Reprinted in *Journal of the Indian Anthropological Society*, (1973), **8**, 131–145.

Mukherjee, D. P. (1972). *Genetic studies in relation to fertility*. Mimeo. report ICMR project.

Mukherjee, D. P. (1984a). Changing patterns of marriage in India and genetical implications. *Human Science*, **33**, 201–219.

Mukherjee, D. P. (1984b). Inbreeding and genetics of quantitative traits in man. In *Human genetics and adaptation*, vol. I, ed. K. C. Malhotra & A. Basu, pp. 533–561. Indian Statistical Institutė of Calcutta.

Mukherjee, D. P., Bhaskar, S. & Lakshmanudu, M. (1974). Studies on inbreeding and its effects in some endogamous populations of Chittoor district, Andhra Pradesh. *Proceedings of the 1st Annual Conference of the Indian Society of Human Genetics*, **27**, p. 17. Also *Bulletin Anthropological Survey of India* (1977), **26**, 10–22.

Mukherjee, D. P. & Chakravarty, S. K. (1977). Aspect of genetical structure in Bengalee Muslims of Cachar. *Bull. Anthropological Survey of India*, **26**, 49–56.

Mukherjee, D. P., Reddy, P. C. & Lakshmanudu, M. (1980). Dermatoglyphic effects of inbreeding. *Journal of the Indian Anthropological Society*, **15**, 67–79.

Murty, J. S. & Jamil, T. (1972). Inbreeding load in the newborn of Hyderabad. *Acta Genetica Medical Gemellol*, **21**, 327–331.

Pai, R. A. (1972). Influence of environmental factors on consanguinity. *Genetics and our health*, ICMR technical report. series, **20**.

Rao, A. P. & Mukherjee, D. P. (1975). Consanguinity and inbreeding effect on fertility, mortality and morbidity in a small population of Tirupati. *Proceedings of the 2nd Annual Conference of the Indian Society of Human Genetics, Calcutta.*

Reddy, B. M. (1981). Population biology of the fishermen of Puri coast. Ph. D. thesis, University of Calcutta.

Reddy, P. C. (1983). Consanguinity and inbreeding effects on fertility mortality and morbidity in the Malas of Chittoor district. *Zeitschrift für Morphologie Anthropologic*, **74**, 45–51.

Roberts, D. F. & Bonne, B. (1973). Reproduction and inbreeding among the Samaritans. *Social Biology*, **20**, 64–70.

Roychoudhury, A. K. (1976). Incidence of inbreeding in different states of India. *Demography India*, **5**, 108–119.

Roychoudhury, A. K. (1980). Consanguineous marriages in Tamil Nadu. *Journal of the Indian Anthropological Society*, **15**, 167–174.

Sabarni, P. & Mukherjee, D. P. (1975). Inbreeding effect on finger print patterns. *Proceedings of 2nd Annual Conference ISHG, Calcutta.*

Sanghvi, L. D. (1966). Inbreeding in India. *Eugenics Quarterly*, **13**, 291–301.

Sanghvi, L. D. & Master, P. A. (1974). Effects of inbreeding on congenital malformation, Abstract. *First Annual Conference ISHG*, p. 17.

Sanghvi, L. D. Varde, D. S. & Master, H. R. (1956). Frequency of consanguineous marriages in twelve endogamous groups in Bombay. *Acta Genetica*, **6**, 41–49.

Schull, W. J., Komatsu, I., Nagano, H. & Yamamoto M. (1968) Hirado: Temporal trends in inbreeding and fertility. *Proceedings of the National Academy of Sciences*, **56**, 671–679.

Schull, W. J. & Neel, J. V. (1972). The effects of parental consanguinity and inbreeding in Hirado, Japan. V, Summary and interpretation. *America Journal of Human Genetics*, **24**, 425–453.

Srinivasan, S. & Mukherjee, D. P. (1976). Inbreeding among some Brahman populations of Tamil Nadu. *Human Heredity*, **26**, 131–136.

Sundar Rao, P. S. & Inbaraj, S. G. (1977). Inbreeding effects on human reproduction in Tamil Nadu of south India. *Annals of Human Genetics*, **41**, 87–98.

Sundar, Rao, P. S. Inbaraj, S. G. & Jesudian, G. (1972). Rural-urban differentials in consanguinity. *Journal of Medical Genetics*, **9**, 174–178.

Veerraju, P. (1973). Inbreeding in coastal Andhra Pradesh. *Proceedings of the International Symposium on Human Genetics*, pp. 309–318. Waltair, India: Andhra University Press.

Veerraju, P. (1978). Consanguinity in tribal communities of Andhra Pradesh, In *Medical genetics in India*, vol. 2, ed. I. C. Verma pp. 157–163.

Vijaykumar, M. & Malhotra, K. C. (1981). Inbreeding and matrimonial distances among the Negroid Siddies of Karnataka (Mimeo). *Technical report Anthropology no.*, **10/81**, Indian Statistical Institute, Calcutta.

8 *Biomedical and immunogenetic variation in isolated populations in India*

SURINDER S. PAPIHA

Although there is a wealth of information available about genetic variation for tribal, rural, semi-urban and urban populations in India (Papiha, 1986), there is as yet little well-documented evidence to suggest how systematic and non-systematic pressures have influenced the present-day genetic pattern.

The majority of Indian populations carry a considerable load of diseases of environmental and especially infectious origin that contribute to childhood mortality. An extreme example is that of the Onge, a negrito tribe from the Andaman Islands. Already much depleted by the ravages of disease imported in the 19th century with the establishment of a penal settlement on the islands (Cappieri, 1947), over the last two to three decades the Onge population has decreased by 80%, and one severe epidemic could wipe out the remaining population.

In India, moreover, there is a wide variety of ecological zones, with their different distributions of vectors of bacterial and parasitic diseases, so that the pattern of human infestation varies from one to another. In spite of extensive efforts, malaria is still periodically experienced, and during the malarial epidemics mortality in certain regions reaches a level 30 times that in normal times, especially in young children (Kirk, 1986). Therefore, it is obvious that selective pressures differing qualitatively and quantitatively are being exercised on different populations in India.

As a measure of the body's defence mechanisms against disease, perhaps the most sensitive is the level of circulating antibodies, or the proteins known as immunoglobulins. The first part of this study assesses the variation in the level of immunogobulins in different tribal, caste and urban populations as an indicator of ecological stress. Secondly it enquires how some of the evolutionary processes may affect the immunoglobulin

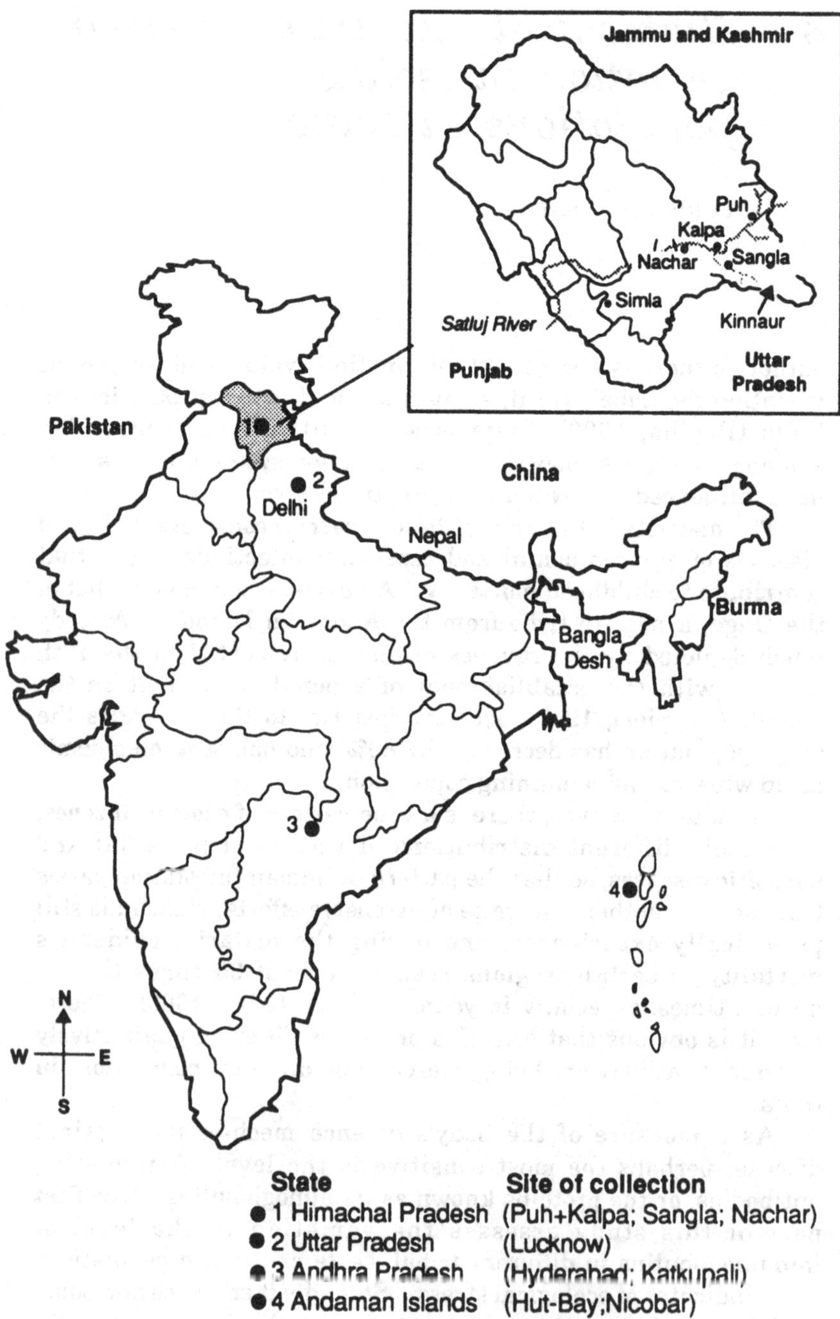

Figure 8.1. Map of India and Himachal Pradesh showing the sites of sample collection.

allotypes and maintain genetic diversity among the local and very diverse populations of India.

The populations sampled (Figure 8.1) for the study of immunoglobulin levels include some who are very different genetically from each other, as follows:

a. regional groups of Kanet (also known as Rajputs) and a low caste Koli population from Kinnaur district in the highland state of Himachal Pradesh in north-west India. The eastern border of this state is shared with Chinese-occupied Tibet. The northern Kanets not only morphologically resemble the mongoloid peoples to the north but culturally also show their influence. In addition to the mountain barriers the great river Sutlej which crosses the valley of Kinnaur restricts the intermixing of the populations from different regions of Kinnaur.
b. the isolated populations of Onge and Nicobarese from the Andaman and Nicobar Islands, situated in the Bay of Bengal, 200 miles from the mainland coast of Madras.
c. the tribe known as Koya from the south Indian state of Andhra Pradesh.
d. healthy individuals from urban Muslim and Hindu populations from south and central India respectively. These were included for comparative purposes.

The levels of immunoglobulins A, M, G and E in these different tribal and urban populations are given in Table 8.1, both as raw data and log values. The IgE values are given in international units (IU/ml) the others as mg/100ml. Figure 8.2 shows the variation of immunoglobulins (raw data) among the different populations. In view of the skewness of the distributions of the raw data in the majority of the samples, highest for IgE, IgM and IgG, logarithmic transformations were applied, and there was a considerable improvement in the normality of their distribution. All statistical comparisons were performed on the logtransformed data.

In the two urban groups the mean immunoglobulin levels are similar and except for IgE there is no significant variation between them. By comparison with European and some other populations (Roberts *et al.*, 1979) their IgG and IgM levels are somewhat high but they may be used as an internal control, a baseline against which the tribal data in this study can be assessed especially since all tests were carried out in the same laboratory using the same procedure.

IgA levels are significantly lower in the Koya tribe and the highland populations of Kanet and Koli; the highest IgA level

Table 8.1. *Immunoglobulin levels in the Indian subpopulations studied*

Population	IgA (mg/100ml)			IgM (mg/100ml)			IgG (mg/100ml)			IgE (IU/ml)			log IgA		log IgM		log IgG		log IgE	
	mean	*n*	*sd*	*mean*	*n*	*sd*	*mean*	*n*	*sd*	*mean*	*n*	*sd*	*mean*	*sd*	*mean*	*sd*	*mean*	*sd*	*mean*	*sd*
Hindu	291.70	59	151.20	191.30	59	89.30	2514.50	58	1080.20	899.00	59	1076.30	2.416	.240	2.237	.201	3.362	.188	2.712	.488
Muslim	282.10	49	106.50	181.40	49	91.20	2575.30	49	977.10	605.30	48	1412.30	2.418	.173	2.207	.214	3.371	.208	2.447	.482
Kota	220.70	101	93.50	172.90	100	83.40	3276.70	102	1113.10	2276.40	100	1744.50	2.302	.203	2.183	.234	3.553	.128	3.251	.302
Onge	247.30	24	74.90	670.80	23	635.70	4470.20	22	1848.10	2953.90	19	1766.50	2.373	.136	2.702	.329	3.607	.220	3.375	.334
Nicobarese	317.80	35	107.10	263.50	35	115.40	2820.10	35	1412.20	4154.40	33	2681.50	2.477	.157	2.384	.182	3.398	.216	3.510	.339
Sangla-Kanet	200.60	95	118.80	172.30	93	63.00	1855.40	94	849.10	431.40	87	504.78	2.229	.270	2.209	.153	3.204	.268	2.364	.502
Nachar-Kanet	175.80	100	90.70	205.00	100	70.00	2217.20	100	631.80	1026.70	97	1125.80	2.203	.189	2.285	.164	3.329	.124	2.777	.490
Pun + Kalpa Kanet	195.80	142	87.50	179.60	142	81.50	1762.30	140	642.30	501.40	135	991.00	2.255	.920	2.210	.203	3.217	.165	2.360	.520
Koli	199.80	61	106.00	217.90	62	86.10	1886.20	62	954.00	584.70	60	947.10	2.230	.275	2.302	.189	3.227	.220	2.410	.555

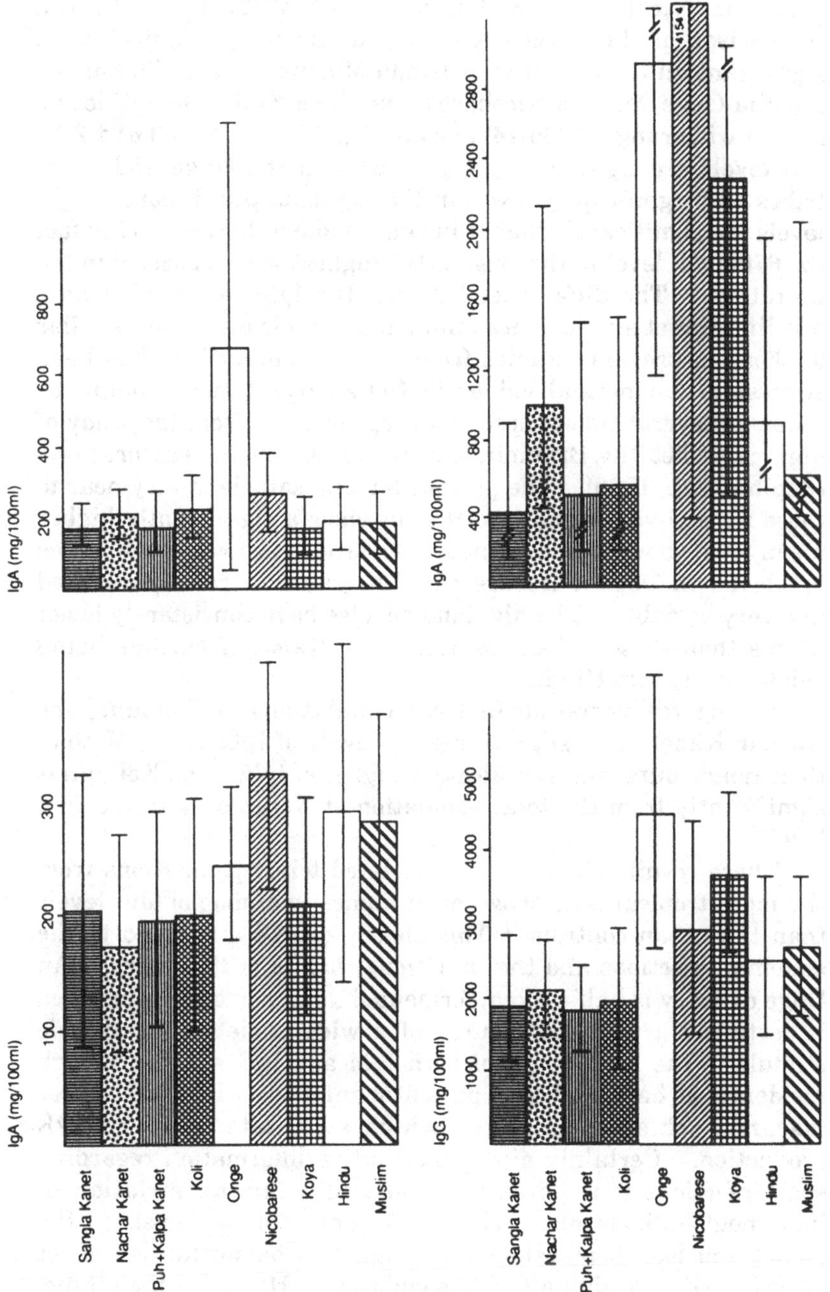

Figure 8.2. Immunoglobulin levels in tribal and non-tribal populations of India.

occurs in Nicobarese (318mg/dl; log 2.48±0.16). Paired comparison of the several populations by t-tests indicated no significant difference between Hindu-Muslim, Hindu-Nicobarese, Muslim-Onge, Muslim-Nicobarese and Koya-Koli. The IgG levels show a wide range (1886-4470mg/dl; log 3.23±0.19 to 3.61±2.1). IgG levels are significantly higher both in the Onge and Koya tribes and significantly lower in the highland populations. IgM levels are significantly higher in Onge and Nicobarese, and in fact the 670mg/dl level in the Onge is the highest so far reported in the literature. The differences between the IgM levels of Hindu, Muslim, Kanet and Koli are minimal and their means are similar to those in previous studies from India. For IgE, it has been suggested that normal values in Indian populations should be taken as several times those in Europeans, and from the study of high caste healthy Brahmin a range of 50-500 IU was proposed (Papiha *et al.*, 1980). The present Muslim sample is very near to those normal values, but the Hindus show a significantly higher mean level though their standard deviations are very high. The IgE levels in Onge, Nicobarese and Koya are extremely elevated and very variable. The highland peoples have consistently lower means than these tribes, all except the Kanet of Nachar being below the Muslim-Hindu.

As regards variation in local populations of Kinnaur, the Nachar Kanet have slightly higher levels of IgG and IgM than their neighbours, and significantly higher of IgE. The Koli differ significantly from the total population of Kanet only in the IgM levels.

These results show that the isolated tribal populations from the more tropical zone show much higher immunoglobulin levels than the urban controls. This clearly can be attributed to the endemic infections and the nutritional habits of the tribes. An Onge delicacy is half-cooked, fermented tortoise meat; such a food is certainly a potential source of a wide variety of antigenic stimuli. The high levels of both IgM and IgE suggest a high incidence of bacterial and parasitic infections in these tribes (ascaris, schistosoma, etc.), which stimulate IgM and IgE production. Certainly more quantitative information regarding such infections is required to account fully for the variation in immunoglobulin levels. The low immunoglobulin levels in the Kanet and Koli populations may suggest a haemodilution effect experienced in a high altitude population. However, the ethnic, regional and intra-regional variations support the possibility of intrinsic factors, and the genetic factor may be one such influence (Billewicz *et al.*, 1974).

Association with genetic markers

Several simple genetic polymorphisms (blood groups, serum proteins and red cell enzymes) were investigated. Several showed different associations with Ig levels, but there was no consistent pattern. In the highland peoples seventeen genetic markers showed association with different levels of immunoglobulins in the Kanet groups. In the tribal populations, in the Onge log IgA levels were associated with PGM1 and EsD alleles while in the Nicobarese log IgA levels were associated with Rhesus and acid phosphatase (AP) alleles. In the Koya the log IgG levels were associated with Fy^a and A^p alleles. In urban populations ten different alleles showed similar associations with immunoglobulin levels. In view of the number of tests the probability levels required correction and when this was done the only association that remained significant was that between log IgG and the Hp polymorphism in Kanets. The interaction of the Hp alleles with Ig levels has been described by other authors (Owen *et al.*, 1964; Nevo & Sutton, 1968; Al-Agidi *et al.*, 1977), all of whose studies indicated higher levels of IgG in Hp2-1 heterozygotes. However in the present study high IgG levels were found in individuals with no detectable haptoglobin, possibly due to the Hp0 allele or possibly to some parasitic infection. This study was not designed to show genetic influence on Ig levels. The population differences in the latter demonstrated here may possibly include some genetic factor, but they must undoubtedly reflect differential stimulation of the immune system by factors of the extreme environmental conditions experienced.

Ecological stress in such isolated populations is certainly great. During their recent microevolution the ancestral populations of these groups must have passed on any selectively advantageous genes that more effectively produce efficient antibodies to meet the differing ecological challenges experienced. If so, and if these challenges are diverse, there may well be differences in the frequencies of such genes, which would give rise to variation in Ig levels among populations of India. If however the challenges are uniform in the different environments, then selection would be directional and the gene frequency more uniform.

If the immunoglobulin genes in Indian populations have passed through such intense selection then one would expect a greater degree of genetic differentiation of alleles at the Ig compared to other loci. To examine this, Ig allotypes were investigated, since the genes for the Ig allotypes are situated

Table 8.2. *Gm haplotype frequencies in populations of India*

Population		Haplotypes					
	G1m	*a(z)*	*a,x(z)*	*f*	*f,a*	*a*	*a(z)*
	G3m	*g*	*g*	*b'*	*b*	*b,s,t*	*b*
Kanet							
(Himachal Pradesh)							
Nachar		0.542	0.216	0.236	-	0.005	-
Sangla		0.402	0.380	0.212	-	0.006	-
Kalpa		0.552	0.225	0.166	0.037	0.019	-
Puh		0.601	0.156	0.068	0.068	0.106	-
Koli		0.335	0.053	0.521	0.020	0.071	-
North-west India							
Brahmin		0.135	0.096	0.596	-	0.019	0.154
Kshatriya		0.211	0.044	0.547	0.039	-	0.159
Vaishya		0.325	0.050	0.550	-	-	0.075
South India							
Hindu		0.384	0.188	0.303	0.048	0.072	-
Brahmin (Madras)		0.328	0.222	0.339	0.023	0.030	0.057
Naicker (Madras)		0.512	0.185	0.255	-	-	0.047
Tibet		0.566	0.099	0.006	0.087	0.242	-

within the loci controlling immunoglobulin structure. This was done in five regional populations of Kinnaur among whom the differences in Ig levels are minimal. These allotype frequencies are put in the perspective of the distribution of Gm haplotype frequency in several other south and north-east Indian populations (Table 8.2).

The distributions of the Gm haplotypes over India show regional patterns. The following haplotypes are particularly interesting as far as the present analysis is concerned: Gm *f,b*, a European haplotype; *fa,b*, a S.E. Asian; and *a,g* and *ax,g*, a combined Asian and European haplotype.

The European *f,b* haplotype frequency is lowest in the Puh Kanets and it increases south-westwards from the border towards the lowland populations of north-west India. The Koli, like the populations of the north-west plains, have the highest *f,b*

Table 8.3. *R matrix of five populations of Kinnaur based on Gm haplotype frequencies*

		1	*2*	*3*	*4*	*5*
1.	Kanet Nachar	0.0108				
2.	Kanet Kalpa	-0.0004	0.0100			
3.	Kanet Sangla	0.0084	-0.0003	0.0552		
4.	Kanet Puh	-0.0242	0.0174	-0.0406	0.1120	
5.	Koli	0.0055	-0.0267	-0.0227	-0.0650	0.1090

Mean diagonal element rii = R_{ST} unweighted = 0.059
weighted = 0.050

haplotype frequency. The south-east Asian haplotype *fa,b* is found in Kanet near the border, and only at very low frequencies in Koli; it is absent or at low frequencies in the other samples. The frequencies of haplotypes *a,g* and *ax,g* are much higher in Kanet than in Koli and other caste populations of the north-west and south. To show genetic relationship and genetic differentiation, these allele frequencies were employed to calculate the conditional kinship matrix R, the principal diagonal element of which describes the overall deviation of the allele frequencies of each population in the array (Table 8.3). The weighted mean of the diagonal elements (R_{ST}) gives the mean genetic heterogeneity of the populations, equivalent to Wright's F_{ST} and Nei's G_{ST}.

In a previous study (Papiha, 1985) where 20 genetic marker loci were used with 56 alleles, the overall weighted genetic differentiation was found to be of the order of 4%, and the several Kanet populations showed clustering or the least differentiation. Using only the present Gm haplotype data in the highland groups, the weighted R_{ST} was 5%, the populations showed greater genetic differentiation, and no regional clustering. For example there is less relationship between Puh and Sangla Kanet than between Koli and Kalpa or Sangla, and the Koli are closer to Nachar than are Kalpa or Puh Kanet. This level of genetic differentiation cannot be accounted for by any homogenising directional selection on immunoglobulin genes; therefore, in the absence of any major environmental heterogeneity in these peoples to promote divergent selection, the alternative and more likely explanation is early differentiation, e.g. by settlement and a founder effect, followed by migration with gene flow from the neighbouring populations.

Table 8.4 *R matrix calculated from Gm haplotype frequency data on Indian populations*

		1	2	3	4	5	6	7	8	9	10	11	12	13	14	15	16
Kanet, Nachar	1	0.061															
Kanet, Kalpa	2	0.057	0.063														
Kanet, Sangla	3	0.065	0.064	0.124													
Kanet, Puh	4	0.039	0.065	0.027	0.145												
Koli	5	0.013	-0.007	-0.014	-0.027	0.059											
Brahmin	6	-0.052	-0.078	-0.042	-0.112	0.032	0.183										
Kshatriya	7	-0.052	-0.074	-0.061	-0.108	0.028	0.157	0.148									
Vaishya	8	-0.016	-0.041	-0.035	-0.074	0.054	0.104	0.094	0.085								
Schd. Caste	9	0.006	0.011	0.002	0.016	-0.026	-0.011	0.001	-0.019	0.079							
Nepalese	10	0.035	0.021	0.002	0.006	0.049	-0.010	-0.004	0.031	-0.013	0.058						
Tibetan	11	0.041	0.059	0.020	0.094	-0.023	-0.120	-0.097	-0.067	0.046	0.013	0.105					
Oraon	12	-0.140	-0.101	-0.136	-0.046	-0.086	-0.053	-0.026	-0.079	-0.052	-0.117	-0.029	0.615				
Hindu, A.P.	13	0.014	0.019	0.027	0.046	0.001	-0.018	-0.034	-0.015	-0.018	0.002	0.007	-0.033	0.035			
Koya	14	-0.127	0.104	-0.123	-0.087	-0.063	0.006	0.028	-0.034	-0.036	-0.097	-0.053	0.491	-0.048	0.414		
Tamil Brahmin	15	0.011	0.006	0.037	-0.006	0.001	0.032	0.015	0.011	-0.003	-0.004	-0.022	-0.072	0.013	-0.055	0.025	
Tamil Naicker	16	0.044	0.039	0.044	0.021	0.010	-0.017	-0.016	0.000	0.017	0.027	0.026	-0.137	0.003	-0.112	0.012	0.041

Mean diagonal element rii = R_{ST} (unweighted) = 0.14

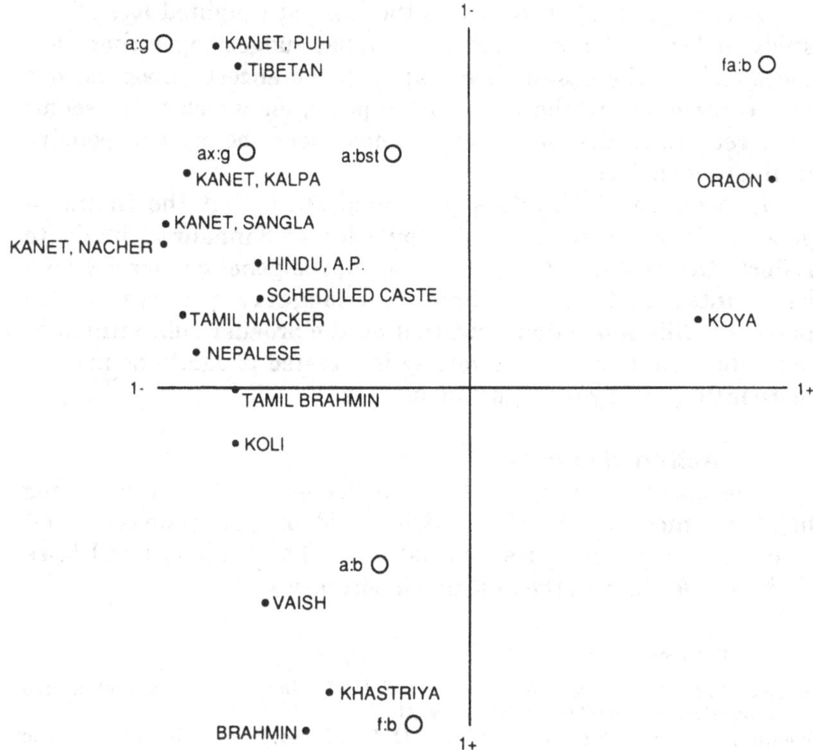

Figure 8.3. Differentiation of Indian populations plotted on first two eigen vectors of R and S matrices for Gm haplotypes.

Using the available Gm haplotype data from 16 widely diverse populations of the Indian subcontinent, genetic affinity and differentiation was also examined using the R matrix (Table 8.4). The R matrix was reduced to an eigenvector plot and was compared with the S matrix to show the allelic contributions (Figure 8.3). The north Indian caste groups show a high caucasoid element with the European haplotype f,b contributing to their differentiation. The Koli of Kinnaur as expected cluster with the low caste groups, while the clustering of the different Kanet samples with other mongoloid groups is attributable to the a,g and ax,g haplotypes. The most highly divergent populations are the tribal populations of Oraon and Koya due to the south-east Asian fa,b haplotype which is the most frequent in both. Although the data are few, it has been suggested that the Gm fa,b haplotype may provide a selective advantage providing resistance to malaria which is epidemic in these populations.

The diagonal element shows the highest weighted R_{ST} of any study of loci so far reported from India, suggesting either that these loci are the best differentiators of the ancestral populations that contributed to the Indian gene pools, for which there seems little reason, or else that these loci have been the most responsive to divergent selection.

In conclusion, the data presented show that the immunogenetic diversity in the local population of Kinnaur is likely to reflect ethnohistory; that, at this level, the genetic diversity may be maintained to some extent by endogamy and systematic pressures like migration; but that on the broader subcontinental level the immunogenetic diversity in isolated populations may be more influenced by the selection process.

Acknowledgement

The author is very grateful to Professor D.F. Roberts for helpful comments and Dr. M.S. Schanfield for typing Gm types and also to several colleagues especially Dr. S.M.S. Chahal and Miss F. Behjati for help in the field and laboratory.

References

Al-Agidi, S. K., Papiha, S. S. & Roberts, D. F. (1977). Immunoglobulin levels in Iraq. *Clinical Experimental Immunology*, **29**, 247–255.

Billewicz, W. Z., McGregor, I. A., Roberts, D. F. & Rowe, D. S. (1974). Family studies of imunoglobulin levels. *Clinical Experimental Immunology*, **16**, 13–21.

Cappieri, M. (1947). II problema dell'Omogeneita razziale degli Andamanesi'. *Rivista di biologia coloniale*, **8**, 59–69.

Kirk, R. L. (1986). Human genetic diversity in south-east Asia and the western Pacific. In *Genetic variation and its maintenance*, ed. D. F. Roberts & G. F. De Stefano, pp. 111–133. Cambridge: Cambridge University Press.

Nevo, S. S. & Sutton, H. E. (1968). Association between responses to typhoid vaccination and human genetic markers. *American Journal of Human Genetics*, **20**, 461–69.

Owen, J. A., Smith, R., Pandanyi, R. & Martin, J. (1964). Serum haptoglobin in disease. *Clinical Science*, **26**, 1–6.

Papiha, S. S. (1985). Genetic structure and microdifferentiation among populations of Kinnaur district, Himachal Pradesh, India. In *Genetic microdifferentiation in human and other animal populations*, ed. Y. R. Ahuya & J. V. Neel, pp. 80–93. New Delhi: Indian Anthropological Association.

Papiha, S. S. (1986). Distribution of some polymorphic red cell enzymes in the Indian subcontinent and their significance: In *Genetic research in India*, ed. I. C. Verma, pp. 71–105. New Delhi: Sagar Printers and Publishers.

Papiha, S. S., Bernal, J. E. & Mehrotra, M. L. (1980). Genetic polymorphism of serum proteins and levels of immunoglobulins and complement levels in a high caste community (Brahmins of Madhya Pradesh). *Japanese Journal of Human Genetics* **25**, 1 0.

Roberts, D. F., Al-Agidi, S. & Vincent, K. (1979). Immunoglobulin levels and genetic polymorphisms in the Sukuma of Tanzania. *Annals of Human Biology*, **6**, 105–109.

9 *Genetic distance analyses in Israeli groups using classical markers and DNA polymorphisms in the β globin gene*

BATSHEVA BONNÉ-TAMIR, AVSHALOM ZOOSSMANN-DISKIN, AHARON TICHER, ARIELLA OPPENHEIM AND SARA NEVO

Introduction

The massive immigration of over a million Jews from all parts of the world into Israel during the last 60 years that is still continuing today (Figure 9.1) provided an opportunity for genetic anthropological studies of the diverse ingathering groups. These led to an appreciation of how heterogeneous is each of these communities, and how much they differ genetically from each other and their previous host populations (Goldschmidt, 1963; Ramot *et al.*, 1973; Bonné-Tamir *et al.*, 1978, 1979). In these studies over the last 28 years various genetic polymorphisms, including red cell and leucocyte (HLA) antigens, serum protein groups and red cell enzyme systems, were used to construct the genetic profiles of each community.

While many studies recognised some degree of affinity between the major Jewish populations (Bonné-Tamir *et al.*, 1979; Karlin *et al.*, 1979; Kobylianksy *et al.*, 1982), the amount of admixture between them and their neighbours remained controversial. Discrepancies in results are due in part to the use of different loci; for example up to 100% admixture of Ashkenazim with other East European peoples was estimated when using the ABO alleles but 0% admixture when the HLA alleles were utilised (Cavalli-Sforza & Carmelli, 1979; Motulsky, 1980). Different methods in deriving admixture rates also yielded contradictory results (Morton *et al.*, 1982; Wijsman, 1984).

A new era in the study of genetic variation began in the 1980s when methods were developed to study polymorphisms directly in the DNA molecules (Botstein *et al.*, 1980). Like colleagues who began to apply analysis of restriction fragment length

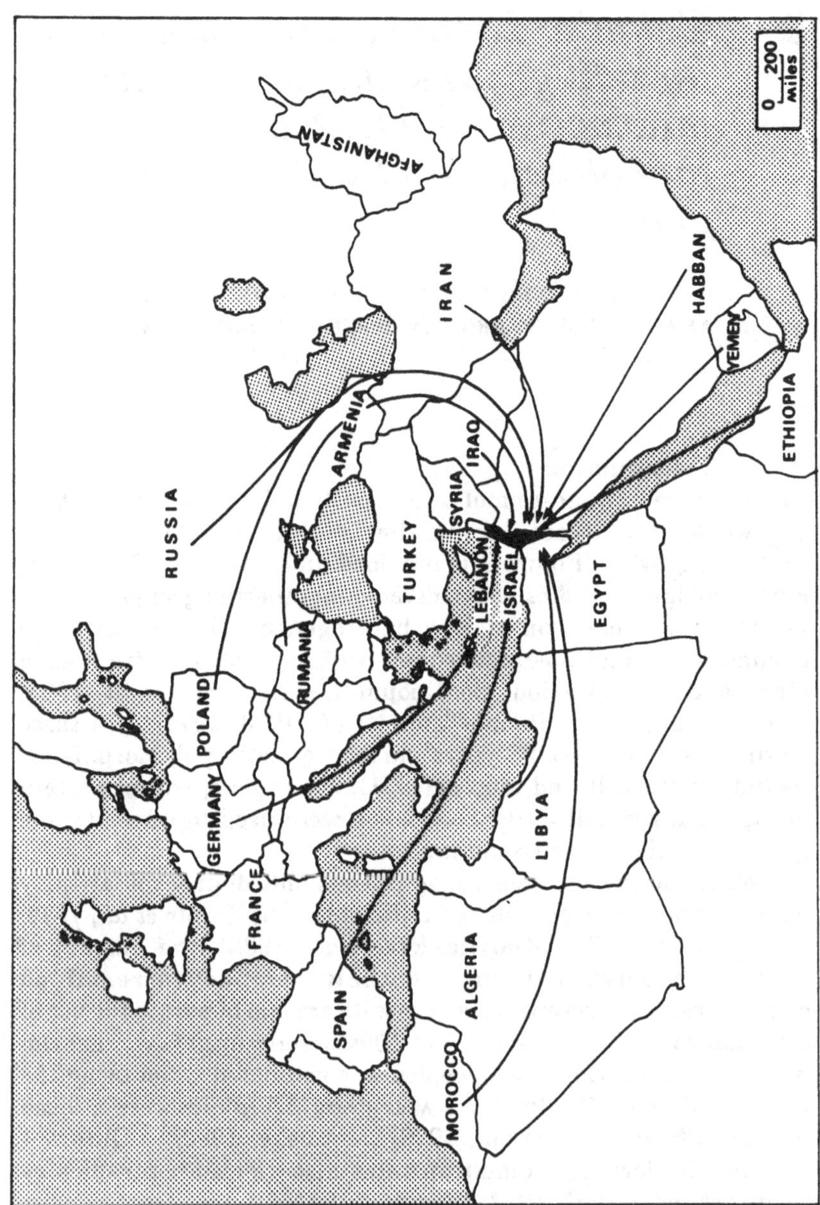

Figure 9.1. Recent Jewish immigrants into Israel.

polymorphisms to population studies (Cavalli-Sforza *et al.*, 1986; Summers, 1987) we used restriction endonucleases and DNA probes to expand the range of informative polymorphisms in the characterisation of the diverse Israeli ethnic groups (Bonné-Tamir *et al.*, 1986; Hakim *et al.*, 1990; Bonné-Tamir *et al.*, 1992). This new and vast resource of genetic markers will supplement those characterised by serological and electrophoretic methods. It will identify specific alleles or new patterns which are expected to provide a clearer understanding of the origins of the groups and the relationships between them.

Results are reported here of an investigation of a nuclear DNA polymorphism in the 5' β globin gene cluster in several Israeli groups and comparison of them with analyses based on three types of classical polymorphism.

The β globin gene cluster

Structure

The human β globin gene family is located on the short arm of chromosome 11 at position 11p15.5 (Grzeschik & Kazazian, 1985). By application of two restriction enzymes (Hinc II and Hind III) and three DNA probes, five polymorphic restriction sites spanning 32kb in the 5' region of the β globin gene complex are detected (Figure 9.2). The presence or absence of the polymorphic site is expressed by the + or - symbol, respectively. Figure 9.2 depicts

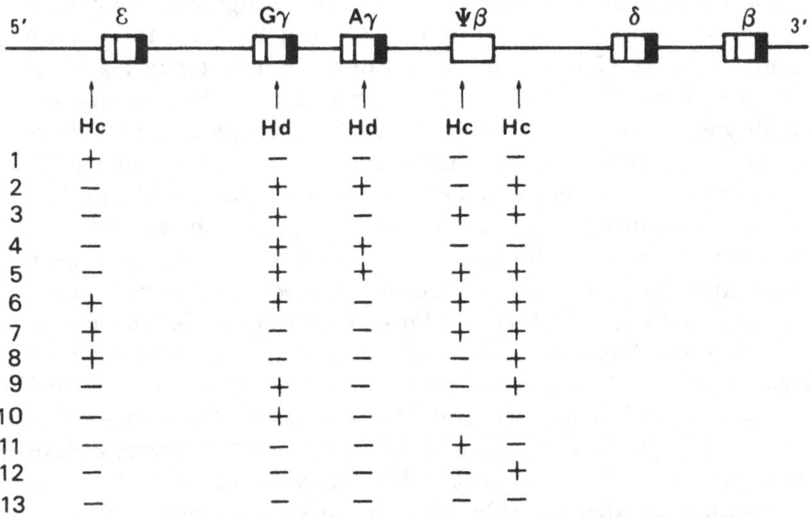

Figure 9.2. Haplotypes in the 5' region of the β globin gene complex.

the positions of the variant sites relative to the genes in the cluster and thirteen haplotypes (combinations of the five sites) derived from them.

These sites were found to be in strong linkage disequilibrium in European, Asian and African populations and are separated from the β globin gene by a recombination hot spot (Antonarakis *et al.*, 1982). Wainscoat *et al.* (1986) pioneered the use of nuclear DNA variations in this cluster to study evolutionary relationships among several racial groups, while Long *et al.* (1990) attempted to determine phylogenetic relationships of the β globin haplotypes themselves.

Haplotypes

On the basis of the five sites described above, 20 out of 32 possible haplotypes have been observed to date in more than 30 populations throughout the world, but few of these reach frequencies greater than 5% in any population (Wainscoat *et al.*, 1986; Long *et al.*, 1990 and see also Appendix A).

DNA samples from members of nuclear families belonging to six different Israeli communities were extracted from buffy coats; digestion, electrophoresis, blotting, hybridisation and autoradiography were carried out using standard methods (Sambrook *et al.*, 1989). A total of 303 chromosomes bearing normal β^A genes were analysed and haplotypes were determined by comparing parents with offspring.

In Table 9.1, showing the samples studied and the frequencies of the haplotypes observed, the haplotype designated number 1 (+ - - - -) which is the most common in all non-African populations is also the most frequent in the Israeli groups including Ethiopian Jews. From the histogram of frequencies of 3 haplotypes in 17 populations (Figure 9.3), for haplotype no.1 there seems to be a cline of decreasing frequency in Eurasia from east to west: from 81% among Chinese to about half (43%) in the British and only 9% among Nigerians and 3% in Benin. The second most common haplotype designated no. 2 (- + + - +) does not show a clear cline but there are variations in frequency levels which correlate with geographical regions; thus frequencies in the Far East (China, Polynesia, Thailand) range from 8% to 11%, while West European frequencies range from 31% (Germans) to 40% (British). Middle-eastern and Mediterranean frequencies are intermediate, from 14% among Greeks and Israeli Arabs to 21% in Irani Jews and 27% in Turks. Haplotype no.12 (- - - - +) is predominantly African; only a few chromosomes bearing this type were found among Germans (6%), Greeks (2%) and in three Jewish

Table 9.1. ***Frequencies of β globin haplotypes in Israeli communities***

Population	*Sample size*	*1*	*2*	*3*	*4*	*5*	*6*	*7*	*8*	*9*	*10*	*11*	*12*	*13*
Polish J.	97	53.6	18.6	13.4	5.1	2.1	0	1.0	0	0	0	2.1	3.1	1.0
Moroccan J.	39	71.8	15.4	2.6	10.3	0	0	0	0	0	0	0	0	0
Irani J.	41	61.0	24.4	9.8	0	0	4.9	0	0	0	0	0	0	0
Yemenite J.	48	70.8	10.4	4.2	2.1	0	0	2.1	0	0	6.3	0	2.1	2.1
Habbanite J.	20	90.0	5.0	5.0	0	0	0	0	0	0	0	0	0	0
Ethiopian J.	22	40.9	13.6	13.6	0	0	0	0	9.1	13.6	4.5	0	4.5	0
Israeli Arab	36	50.0	13.9	22.2	5.6	0	2.8	0	2.8	2.8	0	0	0	0
	303													

For haplotype description see Figure 9.2

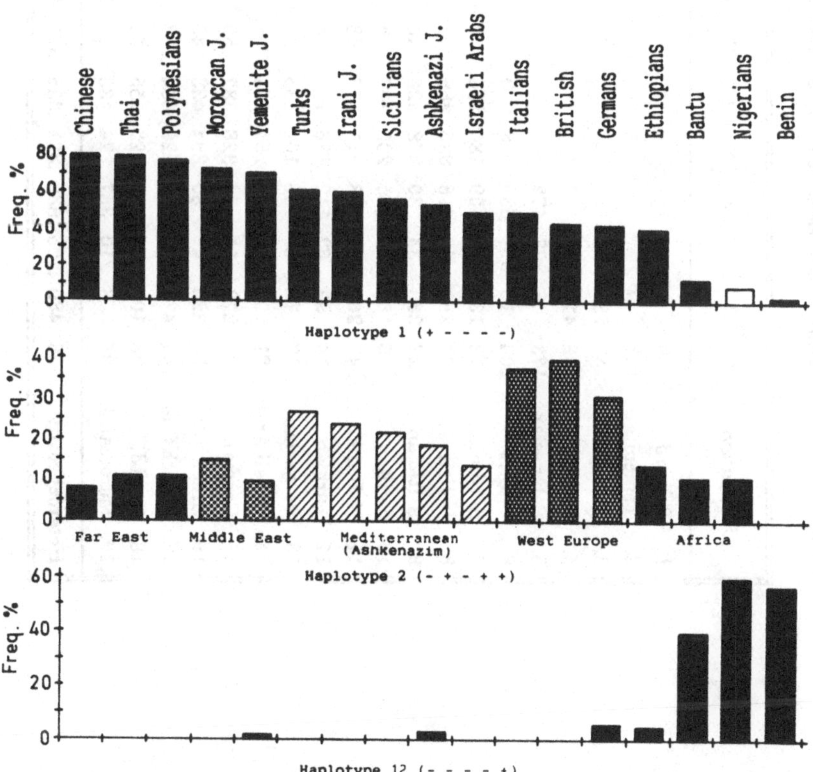

Figure 9.3. Haplotype frequencies in 17 populations.

Table 9.2. *Matrix of genetic distances based on β globin haplotypes (×10³)*

	Population	*1*	*2*	*3*	*4*	*5*	*6*	*7*	*8*	*9*	*10*	*11*	*12*	*13*	*14*	*15*	*16*	*17*	*18*	*19*
1.	Polish J.	0																		
2.	Yemenite J.	131	0																	
3.	Ethiopian J.	47	236	0																
4.	British	99	419	90	0															
5.	Polynesian	173	6	277	472	0														
6.	Chinese	204	16	327	539	8	0													
7.	Sardinian	12	135	44	139	180	204	0												
8.	Indian	37	232	48	18	335	389	59	0											
9.	S. Italian	22	233	33	30	278	326	43	5	0										
10.	Sicilian	6	162	30	85	205	237	8	27	15	0									
11.	N. Italian	47	269	64	18	311	367	88	12	10	44	0								
12.	Turkish	14	99	62	112	126	160	35	53	31	21	44	0							
13.	Bulgarian	17	78	80	136	104	136	38	68	45	29	59	2	0						
14.	Israeli Arab	21	203	34	104	254	284	7	39	31	8	75	52	61	0					
15.	German	33	272	51	24	328	383	59	5	8	31	15	55	67	44	0				
16.	Greek	55	326	49	80	884	433	43	37	42	36	84	107	124	17	40	0			
17.	French Can.	141	476	123	6	529	604	193	41	55	126	30	145	172	153	50	126	0		
18.	Irani J.	28	102	78	122	122	155	51	64	42	33	49	3	5	69	72	131	151	0	
19.	Moroccan J.	96	8	210	357	24	33	97	229	190	123	228	77	56	155	221	263	421	80	0
Heterozygosity		.656	.480	.765	.630	.371	.335	.581	.686	.636	.587	.596	.536	.535	.676	.687	.715	.599	.556	.450

communities: Yemenite Jews (2%), Polish Jews (3%) and Ethiopian Jews (5%).

In addition to the common types, the Israeli samples contain several rare haplotypes found in only a few populations. A unique haplotype no. 11 (- - - + -) which had not been observed before, was identified on two chromosomes in the Polish Jewish sample (Table 9.1).

Genetic relationships between populations based on β globin haplotype frequencies

The haplotypes observed in the Israeli groups were used in the computation of genetic distances among them, and between each of them and a series of other populations. The data sources are listed in Appendix A. The approach suggested by Nei and Tajima (1981) was adopted to compute the distances by using a formula which takes into account the haplotype frequencies and the number of restriction site differences between them, but excluding the sample size factor. The matrix of the genetic distances obtained is shown in Table 9.2.

Among the smallest distances are those between Turks and Bulgarians, British and French Canadians, Chinese and Polynesians. Not surprising also, are the large distances between the Chinese and the British or French Canadians.

Based on the beta globin haplotypes the closest to Polish Jews are Mediterranean populations such as Sicilians and Sardinians, and to Irani Jews are Turks and Bulgarians. Yemenite and Moroccan Jews show proximity to Far Easterners, a situation probably due to the very high frequency of haplotype no. 1 in all of them (Figure 9.2). It is interesting that similarity in frequencies between Yemenite Jews and Chinese has been recently reported for two other nuclear DNA polymorphisms (Kidd *et al.*, 1992). Ethiopian Jews also resemble Mediterranean populations (South Italians, Sicilians and Israeli Arabs) in their beta globin haplotypes, but are distant from Yemenite Jews, a fact pointed out previously regarding other genetic markers (Hakim *et al.*, 1990).

The extent of variation at this locus in each population is shown by the heterozygosity indices (Table 9.2). Chinese and Polynesians show the lowest values (0.335 and 0.371) followed by Moroccan and Yemenite Jews (0.450 and 0.480). In contrast, the small sample of Ethiopian Jews has the highest value (0.765). It is interesting that similar high estimates of heterogeneity among Ethiopian Jews, compared with other populations, was manifested in their mitochondrial DNA types, in which Moroccan and

Table 9.3. *Matrix of genetic distances based on blood groups* ($\times 10^3$)

	Population	*1*	*2*	*3*	*4*	*5*	*6*	*7*	*8*	*9*	*10*	*11*	*12*	*13*	*14*	*15*	*16*	*17*	*18*
1.	British	0																	
2.	French	12	0																
3.	German	34	39	0															
4.	Greek	134	110	148	0														
5.	N. Italian	50	39	52	45	0													
6.	S. Italian	105	39	114	34	33	0												
7.	Sicilian	69	59	97	75	26	64	0											
8.	Sardinian	230	194	210	119	147	143	184	0										
9.	Polish J.	132	126	140	89	75	97	96	217	0									
10.	Moroccan J.	245	234	240	138	136	144	146	235	81	0								
11.	Irani J.	313	318	312	167	200	211	230	298	106	183	0							
12.	Yemenite J.	313	322	309	240	259	284	287	176	274	243	408	0						
13.	Ethiopian	781	776	893	730	738	759	695	869	590	551	765	704	0					
14.	Israeli Arab	305	330	239	272	234	265	294	368	164	236	314	296	748	0				
15.	Chinese	1330	1318	1159	1224	1218	1076	1369	1123	1259	1222	1379	1378	2154	1237	0			
16.	Polynesian	1268	1249	1178	1208	1241	1110	1399	1178	1266	1324	1497	1300	1865	1309	359	0		
17.	Indian	344	305	296	230	214	214	213	245	161	128	254	390	772	278	967	1201	0	
18.	Iraqi J.	333	297	328	126	209	143	273	311	172	213	214	493	834	330	1133	1187	259	0

Yemenite Jews also show relatively low estimates of diversity (Ritte *et al.*,1992).

Genetic distances based on 'classical' markers

The contribution of the 5' β globin haplotypes to the genetic profile and relationships among the populations investigated, by comparison with distance analyses based on other marker systems seemed relevant. Several previous studies dealt with genetic distance analyses between Israeli communities using classical markers (Bonné-Tamir *et al.*, 1978, 1979; Karlin *et al.*, 1979; Kobyliansky *et al.*, 1982; Morton *et al.*, 1982) but they cannot be compared fully with the present analysis since the samples are not the same. Moreover, since these studies were done 10-15 years ago, additional data including more loci have been accumulated, which can provide a more thorough and accurate genetic portrait of these groups.

Gene frequency data comprising 18 serological and electrophoretic loci with 67 alleles, including blood group, protein and enzyme systems and HLA antigens, were used to compute genetic distances between 19 populations studied for the polymorphisms of the 5' β globin gene cluster. The data sources are listed in Appendix B. Distance matrices using the method of Sanghvi (1953) were computed separately for the three types of 'classical' polymorphisms and dendrograms were constructed according to the UPGMA method (Sneath & Sokal, 1973).

These dendrograms are intended only to clarify possible affinities as reflected in the clustering of groups due to similar frequencies and are not attempts to reconstruct phylogenetic relationships (see chapter 12).

a. Blood-groups

Genetic distances for Jewish populations computed for six unlinked loci (ABO, MNS, Rh, Kell, Kidd and P) using 21 alleles and representing at least 5 different chromosomes (Table 9.3) show the shortest distance to be between Polish Jews and north Italians, followed by the former and Moroccan Jews. The estimates of the distances are still several times larger than those encountered among the north-west European populations. Of all pairs of the Jewish populations in the sample, Polish Jews are closest to Irani Jews and Moroccan Jews. A dendrogram based on these distances yields four distinguishable clusters (Figure 9.4): a) North-west European (British, French, German), b) Mediterranean (north and south Italian, Greek and Sicilian), c) Polish, Moroccan and Irani Jews, and d) Far Easterners (Chinese and Polynesian).

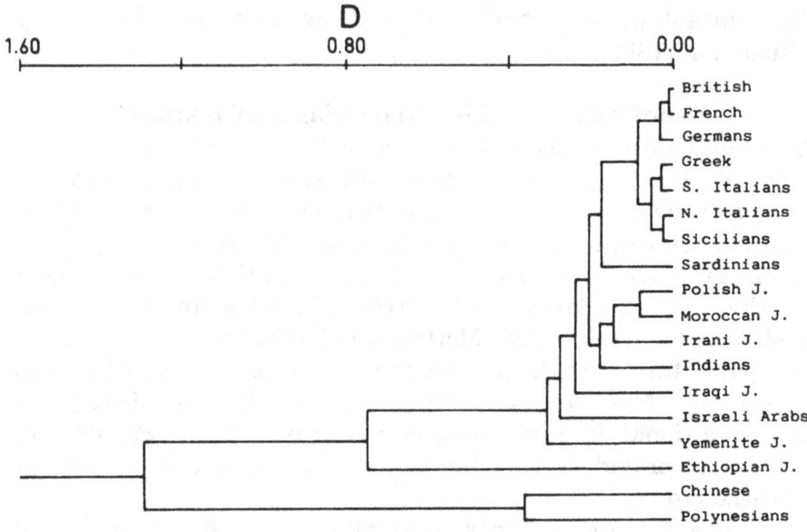

Figure 9.4. Dendrogram of genetic distance based on blood group loci.

b. Proteins

Estimates of genetic distances among the same groups were based on 10 serum protein and red-cell enzyme loci (Hp, Gc, CHE1, ACP1, PGD, PGM1, AK1, ADA, EsD, and GPT1) with a total of 23 alleles (Table 9.4). A basically similar pattern emerges; close populations to Polish Jews and Irani Jews are the Mediterraneans (Italian, Sicilian, Greek); the shortest distance to Moroccan Jews is from the south Italians; Yemenite Jews seem closest to Israeli Arabs. The dendrogram (Figure 9.5) shows clustering of these groups.

c. Leucocyte antigens (HLA)

Twenty three alleles at the HLA A and B loci were used to compute genetic distances (Table 9.5), depicted in the dendrogram (Figure 9.6). These two very polymorphic loci give a somewhat different perspective on the relationships among the same group of populations. Still, the 3 closest groups to Ashkenazi Jews are Mediterranean populations (Turkish, Sicilian and north Italian); so are those to Iraqi Jews (Sicilian, north Italian and Moroccan Jews). The Sicilians are also closest to Moroccan Jews. Analysis of four HLA loci, A, B, C and DR, with a total of 55 alleles but only in 10 populations, yields essentially the same structural relationship.

Table 9.4. *Matrix of genetic distances based on protein loci* ($\times 10^3$)

	Population	*1*	*2*	*3*	*4*	*5*	*6*	*7*	*8*	*9*	*10*	*11*	*12*	*13*	*14*	*15*	*16*	*17*	*18*	*19*	*20*
1.	British	0																			
2.	French	24	0																		
3.	German	13	27	0																	
4.	Greek	101	75	73	0																
5.	N. Italian	39	42	19	42	0															
6.	S. Italian	51	30	42	51	27	0														
7.	Sicilian	56	68	50	92	33	46	0													
8.	Sardinian	54	73	49	139	60	71	57	0												
9.	Polish J.	76	79	55	108	50	60	55	82	0											
10.	Moroccan J.	233	173	189	149	162	109	180	256	160	0										
11.	Irani J.	131	113	118	57	80	70	117	159	92	180	0									
12.	Yemenite J.	354	319	307	245	271	207	250	252	240	264	251	0								
13.	Ethiopian	698	601	664	611	650	533	645	552	606	609	636	247	0							
14.	Israeli Arab	161	144	148	103	115	55	109	155	144	135	129	118	461	0						
15.	Chinese	413	364	422	396	378	313	410	404	497	476	483	519	713	280	0					
16.	Polynesian	614	543	589	527	533	446	586	524	656	563	617	468	647	362	168	0				
17.	Indian	318	299	286	248	250	219	272	393	193	217	208	392	844	224	467	704	0			
18.	Bulgarian	204	194	159	241	144	150	143	166	117	251	221	239	551	217	520	642	311	0		
19.	Turkish	159	136	127	136	120	115	156	226	80	145	100	347	700	217	622	819	198	183	0	
20.	Iraqi J.	199	192	163	74	92	114	134	205	103	193	49	241	689	133	473	618	162	230	142	0

Table 9.5. *Matrix of genetic distances based on HLA A and B loci ($\times 10^3$)*

	Population	*1*	*2*	*3*	*4*	*5*	*6*	*7*	*8*	*9*	*10*	*11*	*12*	*13*	*14*	*15*	*16*
1.	N. Italian	0															
2.	S. Italian	65	0														
3.	Sicilian	95	79	0													
4.	English	576	493	409	0												
5.	French	287	254	177	197	0											
6.	German	238	240	189	168	77	0										
7.	Ashkenazim	299	342	293	629	353	352	0									
8.	Indian	531	563	488	770	503	570	649	0								
9.	Chinese	781	894	723	60	782	755	829	627	0							
10.	Ethiopian J.	863	663	670	944	780	869	832	934	1250	0						
11.	Sardinian	349	363	366	908	656	654	574	876	920	877	0					
12.	Yemenite J.	398	326	340	891	563	577	532	765	1173	735	668	0				
13.	Moroccan J.	352	327	224	409	278	267	376	684	991	788	837	523	0			
14.	Iraqi J.	262	340	231	578	338	345	488	397	900	1026	698	561	274	0		
15.	Israeli Arab	266	271	230	566	288	313	348	391	774	650	602	404	344	360	0	
16.	Turkish	142	139	152	545	261	242	266	409	670	695	579	337	314	305	175	0

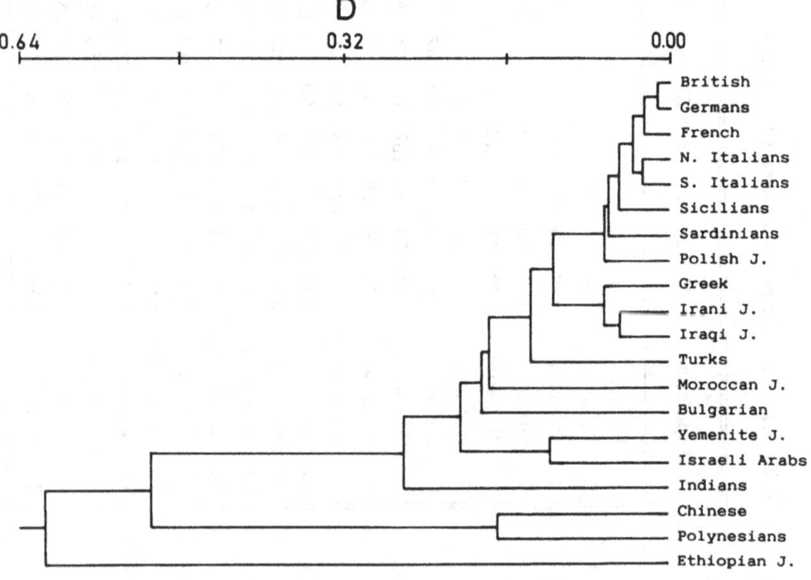

Figure 9.5. Dendrogram of genetic distance based on serum protein and red cell enzyme loci.

d. All markers combined

Combining the frequencies of the classical markers with those of the β globin haplotypes to compute genetic distances among the same population sample gives the matrix shown in Table 9.6. This matrix is based on 17 polymorphic loci (P and Kidd blood groups

Table 9.6. *Matrix of genetic distances based on all systems ($\times 10^3$) including β globin*

	Population	*1*	*2*	*3*	*4*	*5*	*6*	*7*	*8*	*9*	*10*	*11*	*12*	*13*	*14*	*15*	*16*
1.	British	0															
2.	French	261	0														
3.	German	362	351	0													
4.	N. Italian	722	402	521	0												
5.	S. Italian	734	461	655	213	0											
6.	Sicilian	646	428	592	252	221	0										
7.	Sardinian	1363	1174	1206	775	717	641	0									
8.	Polish J.	1016	803	689	636	617	593	979	0								
9.	Moroccan J.	1187	1066	880	945	890	823	1540	821	0							
10.	Irani & Iraqi	1140	869	1003	589	647	618	1264	855	833	0						
11.	Yemenite J.	1957	1646	1561	1288	1054	1150	1359	1251	1207	1463	0					
12.	Ethiopian J.	2762	2505	2683	2557	2151	2307	2442	2320	2399	2705	1901	0				
13.	Israeli Arab	1212	983	1007	857	726	678	1123	807	945	873	1192	1841	0			
14.	Chinese	2895	2561	2606	2403	2274	2289	2410	2473	2527	2460	3000	3797	2475	0		
15.	Indian	1507	1229	1333	1126	1043	1100	1641	1167	1309	985	1840	2844	1070	2096	0	
16.	Turkish	892	573	676	386	393	425	1192	616	869	648	1346	2478	805	2241	916	0

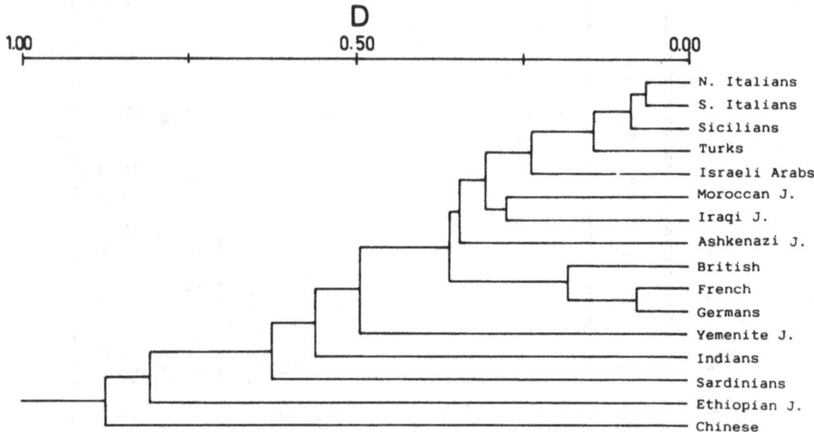

Figure 9.6. Dendrogram of genetic distance based on HLA A and B loci.

are not included) in 16 groups, due to lack of data on all the markers in the other groups. It is astounding that although the 17 gene loci were studied in samples containing different individuals from each population, at different times and locations, and by different investigators, a similar underlying and distinct pattern of population affinities becomes visible (Figure 9.7). While the smallest distances are not surprisingly those between north and south Italians or south Italians and Sicilians, those closest to Polish Jews and Oriental Jews (Iraqi and Irani Jews combined) are, in slightly different order, the same four non-Jewish Mediterranean populations (Sicilian, Turkish, south and north Italian).

Discussion and concluding remarks

The six ethnic groups studied here for the new nuclear DNA polymorphism do not represent the complete Israeli scene of communities. The choice of populations included in the distance calculations was dependent on those that had been investigated for the β globin polymorphisms. The resulting excessive representation of several Italian groups (north, south, Sicily and Sardinia) on the one hand and lack of non-Jewish East-European, North-African and Middle-Eastern populations on the other, may have introduced bias into the comparisons and hence caution is needed in drawing conclusions. At the same time however, these comparisons are probably the most extensive yet performed from the point of view of the number of loci and alleles examined.

Several studies have dealt with the issue of discordant results between distance estimates when based on blood group or protein

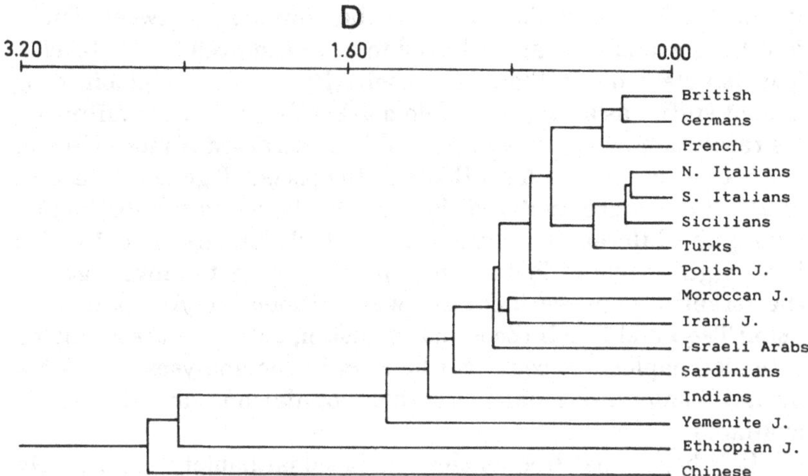

Figure 9.7. Dendrogram of genetic distance based on combined markers.

and HLA loci (Ryman *et al.*, 1983; McLellan *et al.*, 1984 and others). No such discrepancy was obvious in this study. Instead a generally similar pattern of relatedness among the populations was apparent in all four different categories of genetic markers. There was however no exact conformity in the relationships between the major Jewish communities; Polish Jews were close to Moroccan and Irani Jews in their blood group pattern, less so in their proteins, HLA or β globin pattern; Irani and Iraqi Jews were very close in their proteins but not so close in their blood groups. These observations must be substantiated by further examinations in additional communities before an attempt at a definitive explanation is offered. However what was found to be clearly common to the major Jewish communities is their strong Mediterranean affinity in each system separately and in all systems combined. The Mediterranean basin seems therefore to be the most logical place in which to look for their roots or origin.

How have the β globin results affected the genetic profiles of the populations and the general agreement detected in the classical markers? Yemenite and Ethiopian Jews, who differed from each other in most of their genetic markers and also in their mtDNA types, demonstrated also their separate unique patterns of β globin haplotypes. The Caucasoid-Mediterranean component in the Ethiopian gene-pool, traced in many other genetic polymorphisms, was also reflected in their β globin RFLPs (Zoossmann-Diskin *et al.*, 1991). Other supporting evidence for affinities among specific ethnic populations on the basis of the β globin haplotypes is derived from the similarity between Turks and Bulgarians, who are believed to have acquired Central-Asian Turkish elements in their gene-pool. By itself the β globin gene cannot suffice as a single reliable marker for population affinities. Its resolution power, however, could be increased if more sites in the 5' cluster, e.g. the AVA II site in the pseudo β gene (Wainscoat *et al.*, 1984), were analysed, for this would refine the distinction among the different haplotypes, particularly haplotype no. 1 which has a high frequency in many populations. In fact in investigating the Israeli groups we included two additional polymorphic sites using Taq I and Pvu II (Zoossmann-Diskin, 1990) but the resulting extended haplotypes could not be used in the analyses since only meagre comparative data for other populations are at present available.

It is hoped that frequencies of additional haplotypes, not only in this cluster but also in several other nuclear loci currently being investigated (such as phenylalanine hydroxylase, cystic fibrosis), will provide further insights into the issue of past migrations and admixture and allow population affinities to be inferred on broader and more comprehensive evidence.

Acknowledgements

We are grateful to E. Gazit and Y. Shiloh for sharing their DNA samples with us, and to Dr. C. Kirschmann for providing us with the β globin probes. We also thank Z. Goldwitch and M. Haran for their laboratory assistance.

Appendix A

Sources of data used in β globin haplotype distances

Greek, Benin, Thai: Long, J.C. *et al.* (1990) *Am. J. Phys. Anthrop.*, **81**, 113-130.

British, Polynesian, Nigerian: Wainscoat, J.S. *et al.* (1986) *Nature*, **319**, 491-493.

Bulgarian: Kalaydjieva, L. *et al.* (1989) *J. Med. Genet.*, **26**, 614-618.

Chinese (Guandong): Cheng, T. *et al.* (1984) *Proc. Natl. Acad. Sci. USA*, **81**, 2821-2825; Huang, S. *et al.* (1988) *Hemoglobin*, **12**, 621-628; Liu, V.W.S. *et al.* (1988) *Birth Defects*, **23**, 5A, 87-92.

French (Canada): Kaplan, F. *et al.* (1990) *Am. J. Hum. Genet.*, **46**, 126-132.

German, Greek, Turkish: Dehme, R. *et al.* (1985) *Hum. Genet.*, **71**, 219-222.

North Italian (Po River Delta): Toffoli, C. *et al.* (1988) *Eur. J. Haematol.*, **40**, 410-414.

Sardinian: Pirastu, M. *et al.* (1987) *Proc. Natl. Acad. Sci. USA*, **84**, 2882-2885.

Sicilian: Maggio, A. *et al.* (1986) *Hum. Genet.*, **72**, 229-230.

South Italian (Campania): Carestia *et al.* (1987) *Br. J. Haematol.*, **67**, 231-234.

South-African Bantu: Ramsay, M. *et al.* (1987) *Am. J. Hum. Genet.*, 41, 1132-1144.

Turkish: Antonarakia, S.E. *et al.* (1982) *Proc. Natl. Acad. Sci. USA*, **79**, 137-141; Gonzales-Redondo *et al.* (1989) *Blood*, **73**, 1705-1711; Kutlar, A. *et al.* (1989) *Hemoglobin*, **13**, 7-16.

Indians (Sind and Gujarat): Thein, S.L. *et al.* (1984) *Br. J. Haematol.*, **57**, 271-278.

Appendix B

Sources of data used in serogenetic distance

Most of the data were taken from:

1. Mourant, A.E. *et al.* (1976) *The distribution of the human blood groups and other polymorphisms.* London: Oxford University Press.
2. Tills, D. *et al.* (1983) *The distribution of the human blood groups and other polymorphisms, supplement 1.* London: Oxford University Press.
3. Roychoudhury, A.K. & Nei, M. (1988) *Human polymorphic genes, world distribution.* New York: Oxford University Press.

Other data from:

4. Bodmer, J. *et al.* (1972) *Histocompatibility Testing*, pp. 125-132.
5. Bonne-Tamir, B. *et al.* (1978) *Am.J.Phys.Anthrop.*, **49**, 457-464.
6. Bonne-Tamir, B. *et al.* (1978) *Am. J. Phys. Anthrop.*, **49**, 465-472.
7. Bonne-Tamir, B. *et al.* (1979) In *Genetic diseases among Ashkenazi Jews.* (ed. R.M. Goodman & A.G. Motulsky), New York: Raven Press, pp. 54-76.
8. Bonne-Tamir, B. *et al.* (1987) *Gene Geography*, **1**, 1-8.
9. Cohen, T. *et al.* (1981) *Am. J. Med. Genet.*, **8**, 181-190.
10. Golan, R. *et al.* (1977) *Hum. Hered.*, **27**, 298-304.
11. Gualandri, V. *et al.* (1988) *Gene Geography*, **2**, 37-42.
12. Lahav, M. & Szeinber, A. (1972) *Hum. Hered.*, **22**, 533-538.
13. Levene, C. *et al.* (1984) *Israel J. Med. Sci.*, **20**, 509-518.
14. Nevo, S. (1987) *Hum. Hered.*, **37**, 161-169.
15. Nevo, S. *et al.* (1988) *Am.J.Phys.Anthrop.*, **77**, 183-190.
16. Papiha, S.S. & Nahar, A. (1977) *Hum. Hered.*, **27**, 424-432.
17. Peev, Kh. (1980) *Eksp. Med. Morfol.*, **19**, 79-82.
18. Piazza, A. *et al.* (1989) *Gene Geography*, **3**, 69-139.
19. Piazza, A. *et al.* (1989) *Gene Geography*, **3**, 141-164.
20. Rupcheva, L. & Peev, Kh. (1980) *Eksp. Med. Morfol.*, **19**, 14-6.
21. Szeinberg, A. *et al.* (1972) *Clin. Genet.*, **3**, 123-127.
22. Ulizzi, L. *et al.* (1988) *Gene Geography*, **2**, 141-157.
23. Vergnes, H. *et al.* (1980) *Hum. Hered.*, **30**, 171-180.
24. Weissmann, J. *et al.* (1980) *Z. Rechtsmed*, **85**, 55-61.
25. Bonne-Tamir, B. *et al.* (1978) *Tissue Antigens*, **11**, 235-250.
26. Brautbar, H. *et al.* (1992) In *Genetic diversity among Jews: diseases and markers at the DNA level* (ed. B. Bonné-Tamir, A. & Adam). New York: Oxford University Press.

The reference for each marker in each population is given in the table below. In the populations listed, the samples were taken from the same area unless otherwise indicated by a letter after the reference number; British - from England, French - a general sample or from Paris, German - south-west Germany, Greek - a general sample or from Athens, north Italian - Emilia-Romagna and Po river delta, south Italian - Campania, Indian - Gujarat, Chinese - south China. In cases where there were no data on a specific marker in a specific population, data were taken from related populations.

Population	Blood Groups						HLA	Protein Loci									
	ABO	RH	MNS	P	Kell	Kidd		ADA	Ak1	GPT1	ESD	PGM1	PGD	ACP1	GC	HP	CHE1
Polish Jews	13	13	13	7	13	13	3f	7	7	7	10f	7	7	7	14	7	21f
Moroccan Jews	13	13	13	6	13	13	25	6	6	6	10n	6	6	6	1n	6	21n
Irani Jews	13	13	13	5d	13	13	-	9	9	5d	10	9	9	9	1	9	21
Iraqi Jews	5	5	5	5	5	5	25	5	5	5	10	5	5	5	1	5	21
Yemenite Jews	13	13	13	4	13	13	25	2	4	12	10	4	4	4	1	4	21
Ethiopian Jews	8	8	8	8	8	8	26	8	8	8	8	8	8	8	2o	8	1o
Israeli Arabs	1	1	1	1	1	1	25	2	2	12	10	2	15	15	3	15	1
English	3	1	1	1	1	1	3	1	2	3	16	1	1	1	3	2	1
French	1	1	1a	1	1	1	3	3	2	23j	3	2	23j	1	1	1	23j
Germans	1	1	1	1b	1	1b	3	1	2	2	16	1	1	1	1	1	1
Greeks	1	1	1	1	1	3c	-	3	3	3	3c	3	3	3	2	1	2
N. Italians	1	18	18	1	18	18	19	18	18	18	18	18	18	18	18	18	11q
S. Italians	2	18	18	1e	18	3	19g	18i	18i	18	18e	18	18	18	18	18	1r
Sicilians	1	18	18	1	18	18	19	18	18	18	18	18	18	18	18	18	1r
Sardinians	18	18	18	1	18	1	19	18	18	12k	18	18	18	18	18	18	1s
Indians	3	3	3	3	3	3	3	3	3	3l	3	3	3	3	3p	3	3l
Chinese	3	1	3	1	3	3	3	3	3	3	3	3	3	3	3	1	1t
Polynesians	3	3	3	3	3	3	-	3	3	3	3	3	3	3	3	3	1t
Turks								24	1	3	24	24	24	1	1	1	1
Bulgarians								2	2	20	17	2	2	2	1	1	1

a - the same data as for the German sample; b - Hessen; c - Plati; d - the same data as for Iraqi Jews; e - Puglia; f - Ashkenazi Jews; g - Calabria; i - Abruzzo and Molise; j - Toulouse; k - Israeli Arabs; l - Punjab; n - North African Jews; o - Ethiopian Non-Jews; p - North Indians (Delhi); q - Milano; r - Italians general; s - Lebanese; t - Chinese (Australia).

References

Antonarakis, S. E., Boehm, C. D., Giardina, P. J. V. & Kazazian, H. H. Jr. (1982). Non-random association of polymorphic restriction sites in the ß-globin gene cluster. *Proceedings of the National Academy of Science, USA*, **79**, 137–141.

Bonné-Tamir, B., Ashbel, S. & Kennet, R. (1979). Genetic markers: benign and normal traits of Ashkenazi Jews. In *Genetic diseases among Ashkenazi Jews*, ed. R. M. Goodman & A. G. Motulsky, pp. 54–76. New York: Raven Press.

Bonné-Tamir, B., Bodmer, J. G., Bodmer, W. F., Pickbourne, P., Brautbar, C., Gazit, E., Nevo, S. & Zamir, R. (1978). HLA polymorphism in Israel: an overall comparative analysis. *Tissue Antigens*, **11**, 235–250.

Bonné-Tamir, B., Johnson, M. J., Natali, A., Wallace, D. C. & Cavalli-Sforza, L. L. (1986). Human mitochondrial DNA types in two Israeli populations – a comparative study at the DNA level. *American Journal of Human Genetics*, **38**, 341–351.

Bonné-Tamir, B., Zoossmann-Diskin, A. & Teicher, A. (1992). Genetic diversity among Jews reexamined: preliminary analysis at the DNA level. In *Genetic diversity among Jews: diseases and markers at the DNA level*, ed. B. Bonné-Tamir & A. Adam. New York: Oxford University Press.

Botstein, D., White, R. L., Skolnick, M. & Davis, R. W. (1980). Construction of a genetic linkage map in man using restriction fragment length polymorphisms *American Journal of Human Genetics*, **32**, 314–331.

Cavalli-Sforza, L. L. & Carmelli, D. (1979). The Ashkenazi gene pool: interpretations. In *Genetic diseases among Ashkenazi Jews*, ed. R. M. Goodman & A. G. Motulsky, pp. 93–102. New York: Raven Press.

Cavalli-Sforza, L. L., Kidd, J. R., Kidd, K. K., Bucci, C., Bowcook, A. M., Hewlett, B. S. & Friedlaender, J. S. (1986). DNA markers and genetic variation in the human species. *Cold Spring Harbor Symposium on Quantitative Biology*, **LI**, 411–417.

Goldschmidt, E. (1963). *The genetics of migrant and isolate populations*. New York: The Williams & Wilkins Company.

Hakim, I., Gross, M. & Bonné-Tamir, B. (1990). Genetic relationships between Ethiopian and Yemenite Jews by means of mtDNA polymorphisms. In *Pluridisciplinary approach to human isolates*, ed. A. Chaventre & D. F. Roberts. Paris: INED.

Grzeschik, K. H. & Kazazian, H. H. (1985). Report of the committee on the genetic constitution of chromosomes 10, 11 and 12. *Cytogenetic Cell Genetics*, **40**, 179–203.

Karlin, S., Kenett, R. & Bonné-Tamir, B. (1979). Analysis of biochemical data on Jewish populations. II. Results and interpretations of heterogeneity indices and distance measures with respect to standards. *American Journal of Human Genetics*, **31**, 341–365.

Kidd, K. K., Kidd, J. R., Bonné-Tamir, B. & New, M. (1992). Nuclear DNA polymorphisms and population relationships. In *Genetic diversity among Jews: diseases and markers at the DNA level*, ed. Bonné-Tamir & A. Adam. New York: Oxford University Press.

Kobyliansky, E., Micle, S., Goldschmidt-Nathan, M., Arensburg, B. & Nathan, H. (1982). Jewish populations of the world: genetic likeness and differences. *Annals of Human Biology*, **19**, 1–34.

Long, J. C. Chakravarti, A., Boehm, C. D. Antonarakis, S. & Kazazian, H. H. Jr. (1990). Phylogeny of human ß-globin haplotypes and its implications for recent human evolution. *American Journal of Physical Anthropology*, **81**, 113–130.

McLellan, T., Jorde, L. B. & Skolnick, M. H. (1984). Genetic distances between the Utah Mormons and related populations. *American Journal of Human Genetics*, **36**, 836–857.

Morton, N. E., Kenett, R., Yee, S. & Lew, R. (1982). Bioassay of kinship in populations of Middle Eastern origin and controls. *Current Anthropology*, **23**, 157–162.

Motulsky, A. G., (1980). Ashkenazi Jewish gene pools : admixture, drift and selection. In *Population structure and genetic disorders*, ed. A. W. Eriksson. pp. 353–365. London: Academic Press.

Nei, M. & Tajima, F. (1981). DNA polymorphism detectable by restriction endonucleases. *Genetics*, **97**, 145–163.

Ramot, B., Adam, A., Bonné B., Goodman, R. M. & Szeinberg, A. (1973). Sheba international symposium. *Israeli Journal of Medical Science*, **9**, 1129–1533.

Ritte, U., Neufeld, E. & Bonné-Tamir, B. (1992). Types of mitochondrial DNA among Jews. In *Genetic diversity among Jews: disease and markers at the DNA level*, ed. B. Bonné-Tamir & A. Adam. New York: Oxford University Press.

Ryman, N., Chakraborty, R. & Nei, M. (1983). Differences in the relative distribution of human gene diversity between electrophoretic and red and white cell antigen loci. *Human Heredity*, **33**, 93–102.

Sambrook, J., Fritsch, E. F. & Maniatis, T. (1989). *Molecular cloning: a labratory manual*. New York: Cold Spring Harbor Labratory Press.

Sanghvi, L. D., (1953). Comparison of genetical and morphological methods for a study of biological differences. *American Journal of Physical Anthropology*, **11**, 385–404.

Sneath, P. H. A. & Sokal, R. R., (1973). *Numerical taxonomy*. San Francisco: Freeman.

Summers K. M. (1987). DNA polymorphisms in human population studies: a review. *Annals of Human Biology*, **14**, 203–217

Wainscoat, J. S., Hill, A. V. S., Boyce, A. L., Flint, J., Hernandez, M., Thein, S. L., Old, J. M., Lynch, J. R., Falusi, A. G., Weatherall, D. J. & Clegg, J. B. (1986). Evolutionary relationships of human populations from an analysis of nuclear DNA polymorphisms. *Nature*, **319**, 491–493.

Wainscoat, J. S., Old, J. M., Thein, S. L. & Weatherall, D. J. (1984). A new DNA polymorphism for prenatal diagnosis of β-thalassaemia in Mediterranean populations. *Lancet*, **ii**, 1299–1301.

Wijsman, E. M. (1984). Techniques for estimating genetic admixture and applications to the problem of the origin of the Icelanders and the Ashkenazi Jews. *Human Genetics*, **67**, 441–448.

Zoossmann-Diskin, A. (1990). Polymorphic restriction sites in the β-globin gene cluster: a comparative study in four Jewish communities. M. Sc. thesis, Tel-Aviv University, Israel.

Zoossmann-Diskin, A., Teicher, A., Hakim, I., Goldwitch, Z., Rubinstein, A. & Bonné-Tamir, B. (1991). Genetic affinities of Ethiopian Jews. *Israeli Journal of Medical Science*, **27**, 245–251.

10 *Non-random distribution of Gm haplotypes in northern Siberia*

R. I. SUKERNIK

The investigation of the genetic polymorphisms of the immunoglobulin genetic markers has been for many years the central point in the population surveys carried out in Siberia. The object was to exploit rapidly disappearing opportunities to attain three particular and related aims: (1) to establish the major features of the genetic structure of circumpolar ethnic groups, the least disturbed by recent modernisation and gene flow from outside; (2) to trace their population and regional history and affinities; (3) to search for Gm haplotypes unique to such populations in this area, thus extending knowledge of variation and evolution of these particular loci.

An extended array of Gm allotypic determinants was serologically defined in 31 circumpolar populations belonging to eight ethnic groups (Figure 10.1), including local inland

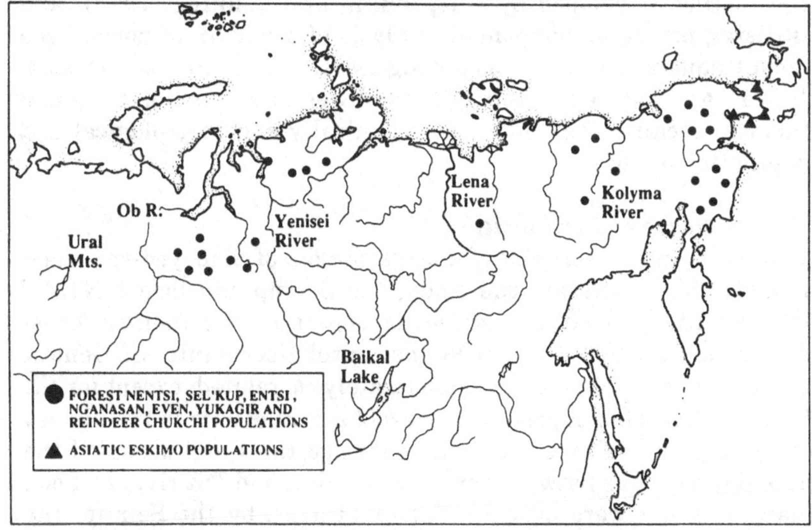

Figure 10.1. Location of samples.

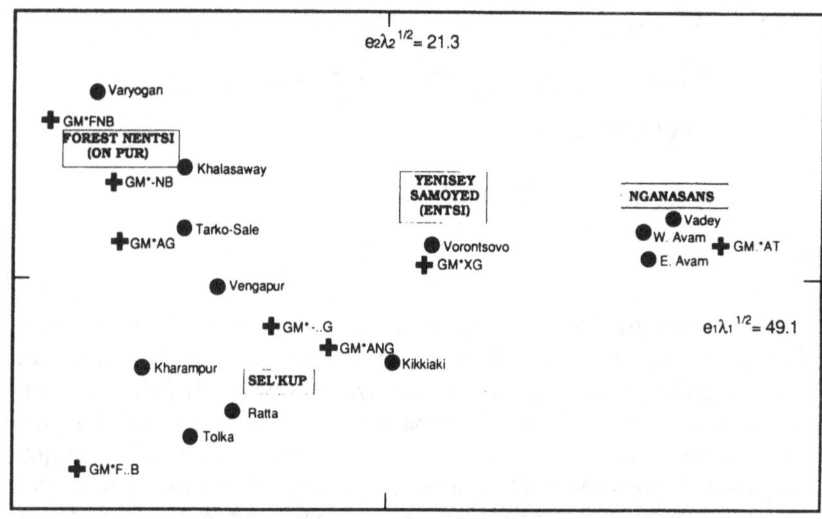

Figure 10.2. Combined plot of 12 Samoyed-speaking populations (●) and 8 GM haplotypes (+).

populations and coastal eskimoan populations. The sample size usually exceeded one half of the census size of the community, and in most cases numbered more than one hundred individuals.

To show the main features of the regional genetic structure, the method developed by Harpending and Jenkins (1973), and utilising principal component analysis of a matrix of normalised gene frequency covariances among local populations, was chosen. It was applied separately for northwestern and northeastern Siberia, because they differ geographically and in geological and population history.

Northwest Siberia

In a two dimensional plot of a large sample of Samoyed-speakers, namely the Nganasan, the Entsi, the Selkup and forest Nentsi (Figure 10.2), the first eigenvector separates the fishing forest Nentsi and the Selkup from Nganasan reindeer hunters. Genetic and geographic distances are moderately correlated, except for the Entsi. This is not suprising in view of recent events in the history of the region. Three hundred years ago the Entsi occupied the boreal forest zone between the lower Yenisei and Taz rivers. Soon many of them were either killed or overrun by the Selkup, the dwellers of the Taiga area, who were then expanding. The remaining Entsi were forced northward into the open tundra, and are currently living in western Taimir, close to the lands of the

Nganasan. Thus local events changed the direction and amount of differential gene movement, and so account for the minor discordance between genetic and geographic distance. The Nganasan groups genetically represent a tight cluster, though considerable geographic distances separate the three current subdivisions. The genetic homogeneity of the present population is to be explained by the cumulative effects of recent fusion of small Nganasan subdivisions, bilineal exogamy, and negligible or no gene flow from nearby Entsi. The second eigenvector reflects the dispersion of isolated Nentsi and Selkup villages varying in size, and scattered over the vast plain and lake area of the West Siberian lowlands.

The contribution of the Gm haplotypes to the variation along each of the axes identifies those that are associated with the respective populations. Forest Nentsi and the Selkup have both the 'European' **f n b* and **f .. b*, as well as the 'Northern mongoloid' **a t*. The latter haplotype is prevalent in the Nganasan, and is the second most frequent haplotype in the Selkup and forest Nentsi and the intermediates, that is the Entsi.

An extremely interesting finding was the deleted **- n b* haplotype, identified in 29 (4%) forest Nentsi. The other deleted haplotype **- .. g* confined to the Selkup does not, however, approach polymorphic frequency. Each haplotype is restricted to within its own tribal boundaries with apparently rare exceptions, thus indicating only occasional gene exchange. Both haplotypes may have originated by unequal crossing-over and are notable for the loss of an Igγ1 gene. The haplotype occurs in the homozygous state in three siblings of mixed Nentsi-Selkup origin. At the DNA level the deletion has been confirmed recently by Southern blot procedures using appropriate restriction enzymes and specific probes.

Northeast Siberia

Essentially less genetic diversity was encountered in aboriginal populations in northeast Siberia (Figure 10.3). An array of three Gm haplotypes, **a g*, **x g*, and **a t*, first noted in the Nganasan is ubiquitously present in the Tungus-speaking Evens, the Yukagirs, and the palaeoasiatic-speaking Chukchi. Asiatic Eskimos show the most impoverished gene profiles, restricted to only two Gm haplotypes, **a g* and **a t*. This pattern, notable by the absence of **x g*, is observed in all unadmixed Eskimoan communities from Siberia to Greenland. Close affinities of the Eskimoan populations are reflected on the plot in Figure 10.3. Most genetic distances between them are very small and poorly correlated with

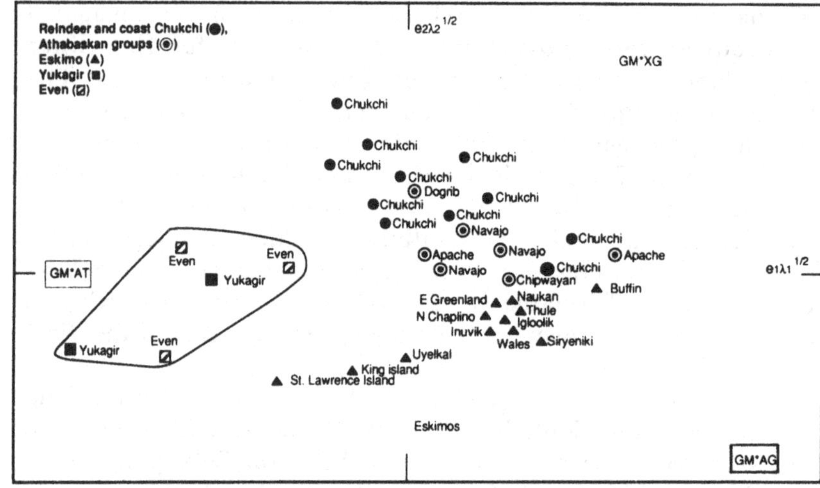

Figure 10.3. Genetic relationship among 32 northeastern Siberian-North American populations.

geographic distance and language affiliation. The most likely explanation is that these populations share a recent common ancestor. This interpretation from this analysis is supported by anthropological, historical and archaeological data. The position of the Saint Lawrence Eskimos is associated with the contribution of the **a t* haplotype occurring at a frequency which is the highest in the total Eskimoan population. The presumed agent responsible for the observed differentiation of the Saint Lawrence Eskimos is a bottleneck effect in their recent history.

A Chukchi-Athapaskan continuum is also apparent. There are only minor differences between these stocks, again in amount of the 'Northern mongoloid' **a t*, which is at somewhat greater frequency in reindeer Chukchi.

The peopling of America

The Siberian data support the idea of three distinct Gm patterns and distributions associated with three main stocks peopling the New World before and after the last glaciation, as in Table 10.1.

The simplest explanation of these comparatively common genetic profiles is that they are the result of three distinct migrations from different Siberian homelands at different times. Perhaps the following events occurred. The mammoth hunters, the first to move south from the Beringia tundra at least 20,000 years ago, had not only the least extreme mongoloid physical phenotype, but also an impoverished Gm pattern, notable by the

Table 10.1. *Gm patterns postulated in 3 main stocks of New World immigrants*

Stock	*Gm* *ag	*xg	*at	*Genetic profile can be readily*
Palaeoindians	+	+	-	derived from Chukchi
Athapaskan Indians	+	+	+	shared with Chukchi
NW Eskimos	+	-	+	shared with Asiatic Eskimos

absence of the 'Northern mongoloid' marker, *Gm*a t*. A second wave of hunters (the Nadene) originated in the boreal forest area somewhere at the junction of Siberia and Alaska approximately 12,000 years ago. That is why the Chukchi who originated in the same or adjacent boreal forest area show close genetic affinity to American Indians.

With respect to the origin of the Eskimos, their unique genetic profile and maritime hunting and gathering culture suggest that the homeland of their Eskimoan ancestors lay far south of the Bering land bridge. Perhaps, they entered the New World by a slightly different route some 9,000 years ago.

In conclusion, the Gm patterns and distributions in holarctic human populations are consistent with the hypothesis that three successive waves migrated across the Bering land bridge from easternmost Siberia during the peopling of the New World.

References

Harpending, H. & Jenkins, T. (1973). Genetic distance among southern African populations. In: *Methods and theories of anthropological genetics*, ed. M. H. Crawford & P. L., Workman, pp. 177–199. Albuquerque: University of New Mexico Press.

11 *Allele frequency estimation*

NORIKAZU YASUDA

The genetic structure of a human population can be described in terms of two principal parameters: the inbreeding coefficient and gene frequency array. The inbreeding coefficient specifies the correlation between uniting gametes, and therefore alleles, in a given gene pool. Allele or gene frequency is a fundamental quantity by which to describe the wide occurrence of genetic polymorphisms in man. Recent advances in molecular biology have confirmed the existence of multiple alleles at a single locus, as well as multiple loci which are linked on the same chromosome.

This paper attempts to unify formulae for estimating allele frequency from population samples, by a counting method. It also presents quantities expressed in terms of allele frequencies, polymorphic information content (PIC) and probability of paternity exclusion (PE). These are of interest in genetic epidemiology.

With desktop computers there is little difficulty in using mathematically quite sophisticated methods such as maximum likelihood, but handy formulae are still indispensable and sound algorithms are required. Many genetic systems can now be treated as codominant, but some are still complicated by dominance. In the following, calculation procedures are given using codominant systems and the ABO-like systems.

Allele frequency estimations

The codominant system

Consider a locus A, with two alleles A_1 and A_2. Let the frequency of A_i be p_i ($i=1$ and 2). The expected frequency of the three genotypes, A_1A_1, A_2A_2 and A_1A_2, in a random mating population, will be p_1^2, p_2^2 and $2p_1p_2$, respectively. By random sampling of n unrelated individuals, among which n_1, n_2 and n_{12} are respectively the number of genotypes A_1A_1, A_2A_2 and A_1A_2 and $n=n_1+n_2+n_{12}$, a gene counting method provides a simple estimate of the allele frequency:-

$p_1 = (2n_1 + n_{12})/2n$ for allele A_1

and $p_2 = (2n_2 + n_{12})/2n$ for allele A_2.

If all individuals are heterozygous, as may happen when the sample size is small, then $p_1 = p_2 = 1/2$.

An extension to a multi-allelic locus is now straightforward and allele frequencies can be expressed as:

$$p_i = (2n_i + n_{ij})/2n \text{ for allele } A_i \ (i \neq j)$$

where n_{ij} is the number of heterozygotes with gene A_i. Recent advances in molecular biology support the view that almost all genetic systems including restriction fragment length polymorphisms and variable number tandem repeat sequences can be treated as codominant systems.

Another extension to linked loci will be more practical, as occurs with RFLP sites located in tandem on the same chromosome. First of all, consider two linked loci A and B, at each of which there are two alleles A_1 and A_2, and B_1 and B_2, respectively. Let the frequencies of the four haplotypes A_1B_1, A_1B_2, A_2B_1 and A_2B_2 be r_{11}, r_{12}, r_{21} and r_{22}, respectively. Then $p_1 = r_{11} + r_{12}$ and $q_1 = r_{11} + r_{21}$ are allele frequencies of A_1 and B_1, respectively. Suppose that n unrelated individuals are samples from a random mating population. Then there are nine possible phenotypes according to the allelic constitution at the two loci. However, the number of possible genotypes is now ten. This disagreement is due to the fact that linkage phase in the double heterozygote is not known in a population sample. Therefore it is necessary to introduce the probability of being in coupling (or repulsion) phase for the doubly heterozygous individual.

By chromosome counting, which also supplies maximum likelihood estimates, are obtained

$$r_{11} = a_1 + bh, \quad r_{12} = a_2 - bh, \quad r_{21} = a_3 - bh \text{ and } r_{22} = a_4 + bh$$

in which the a's and b are shown in terms of the observed numbers (Yasuda, 1978a). The probability, h, that an individual with the doubly heterozygous phenotype $A_1A_2B_1B_2$ possesses genotype A_1B_1/A_2B_2 will be 1/2 when $r_{11}r_{22} = r_{12}r_{21}$, that is when the population is at linkage equilibrium. Then, the estimation procedure becomes independent between two loci. When all individuals are doubly heterozygous, the estimated haplotype frequency is $r_{ij} = 1/4$ irrespective of the size of population. If all individuals are heterozygous at one of the loci, say locus B, the B_1 allele frequency becomes 1/2. After estimating allele frequencies at the A-locus, the haplotype frequencies can be calculated simply by taking the products of two allele frequencies from loci A and B, respectively, as if they are independent of each other.

These algorithms for estimating haplotype frequency when loci are linked and there are two alleles at each locus, have been developed for use with a desktop computer. The number of possible genotypes where all loci are heterozygous increases as 2 raised to the power of the number of loci.

The dominant system

When there are two alleles at a single locus, dominance reduces the number of phenotypes to two. Let A and aa be the dominant and recessive phenotypes, respectively. By definition, there are two possible genotypes, AA and Aa, in a dominant phenotype. Introducing a parameter h as the probability of being homozygous given the dominant phenotype, the frequency of the dominant allele, p_A, is estimated from

$$p_A = n_A(1+h)/2n$$

where n is the size of sample, and n_A is the observed number of the dominant phenotype. In a random mating population, the probability h can be shown in terms of the dominant gene frequency; $h = p_A/(2-p_A)$. Substituting this for the previous equation gives a quadratic equation with p_A, which can be solved to give

$$p_A = 1 - \sqrt{(n-n_A)/n}$$

A logical extension to the multiple allele case is to assume that there is an arbitrary number of codominant alleles and a silent allele. Such examples are the ABO blood group locus with two codominant alleles (A and B) and the recessive gene (O), and the histocompatibility loci (HLA) at which there are many codominant alleles and, for technical reasons, there is almost always one 'silent' allele for undetected antigens.

Let the phenotype symbol designate the observed number and n the sample size. Allele frequency in the ABO system can be estimated by gene counting (Yasuda & Kimura, 1968) as

$$p = \{AB + A(1+h_A)\}/2n \text{ for allele A } \{h_A = p/(p+2r)\}$$
$$q = \{AB + B(1+h_B)\}/2n \text{ for allele B } \{h_B = q/(q+2r)\}$$

and $r = 1-p-q$ for allele O.

In practice, p, q and r are obtained by iteration starting from

$$p = \{(1/2)AB + A + O - \sqrt{O(A+O)}\}/n,$$
$$q = \{(1/2)AB + B + O - \sqrt{O(B+O)}\}/n$$

and $r = 1-p-q$, which already gives an excellent approximation (Yasuda, 1984).

For HLA or other ABO-like systems, the extension of these formulae is straightforward. Instead of the AB-phenotype in the ABO blood groups, is simply substituted the sum of the observed numbers of heterozygotes who possess the allele in question and one of the other detectable alleles or antigens.

The estimation procedure can now be extended to multi-locus problems, as treated in the codominant system. First, suppose that there are two loci, A and B, at each of which there are two alleles one dominant to the other, respectively A and O, and B and O. Using the notation of phenotypes for the observed number and n for the size of sample, the frequencies of four possible haplotypes are given by

$$r_{OO} = \sqrt{O/n},$$
$$r_{AO} = \sqrt{(A+O)/n} - \sqrt{O/n},$$
$$r_{OB} = \sqrt{(B+O)/n} - \sqrt{O/n},$$

and

$$r_{AB} = 1 + \sqrt{O/n} - [\sqrt{(A+O)/n} + \sqrt{(B+O)/n}].$$

Specifically, we are interested in the formula for the AB-haplotype. If a third locus C is then considered with two alleles as in the case of the A or the B locus, the frequency of haplotype ABC can be estimated from

$$r_{ABC} = 1 - \sqrt{O/n} + [\sqrt{(A+O)/n} + \sqrt{(B+O)/n} + \sqrt{(C+O)/n}]$$
$$-[\sqrt{(A+O)^*(B+O)/n} + \sqrt{(A+O)^*(C+O)/n} + \sqrt{(B+O)^*(C+O)/n}]$$

in which, for example, $(A+O)^*(B+O) = AB + A + B + O$ as a symbolical presentation.

Thus, when an arbitrary number of loci is considered, at each of which there are a dominant and a recessive allele, an analytical formula can be derived as a logical extension of this procedure (Yasuda, 1978b).

An algorithm is also available for a two-locus-problem with two ABO-like systems, in each of which there is an arbitrary number of codominant alleles (Yasuda & Tsuji, 1975).

Other useful quantities

Polymorphic information content

One of many important aims in genetic epidemiology is to map disease loci on a specific chromosome. In this regard, linkage analysis with polymorphic markers whose map locations are known is indispensable, and is the more informative if the marker locus shows a high heterozygosity. Botstein *et al.* (1980) introduced a more specific quantity, the polymorphic information content or PIC. The PIC is the probability of establishing linkage phase if grandparents are not examined for the first child. Once phase is established, it is the probability that an offspring is informative for accepting or rejecting linkage (Skolnick, 1986). For a locus with two codominant alleles, the PIC is $2p_1p_2(1-p_1p_2)$, where p_i is the allele frequency ($i=1$ and 2). The PIC for a codominant locus with m alleles can be shown in terms of the power-sum of the allele frequency, namely,

$$PIC = 1 - S_2 + (S_4 - S_2^2)$$

in which $S_k = p_1^k + p_2^k + ... + p_m^k$ and p_i is the frequency of the i-th allele (i = 1,...,m).

Probability of paternity exclusion

This quantity, as the name suggests, is used in forensic medicine, but an interesting application has been made in population surveys of spontaneous mutation rates deduced from the electrophoretic mobility of polypeptides (Neel *et al.*, 1986). It is used to make a corrrection of observed mutation rate for undetected discrepancies between legal and biological parentage.

Formulae are given here in terms of power-sum only for the codominant system and the ABO-like system. For a codominant system

$$PE(m) = 1 - 2S_2 + S_3 - 2S_2^2 + 2S_4 + 3S_2S_3 - 3S_5 \quad \text{(Komatsu, 1952a)}$$

and, for an ABO-like system

$$P(m) = PE(m-1) - \{T_2^2 - T_4\} - r\{r(2-r) + (4-r-r^2)T_2 - 2(1+r)T_3 - 2T_4 + T_2^2\}$$

where $S_k = \sum_{i=1}^{m} p_i^k$; $T_k = \sum_{i=1}^{m-1} p_i^k$; p_i(i = 1,...) is the frequency of the i-th codominant allele; and r is the recessive frequency. When the frequency of the recessive gene approaches zero, so that the system has (m−1) codominant alleles, the limiting probability of paternity exclusion in such a system is always smaller than in the corresponding codominant system. The fact with m = 3 was first reported by Komatsu (1952b).

Summary

There is no doubt that allele frequencies are the fundamental parameter for describing human populations. The counting method from population samples is simple for estimating allele frequency. New algorithms and formulae are developed for estimation of frequency of a haplotype or a set of tandemly linked loci. It is also shown that other quantities such as PIC and PE which are of interest in genetic epidemiology, though little use has yet been made of them in isolate studies, can be presented in terms of power-sum of allele frequency.

References

Botstein, D., White, R. L., Skolnick, M. & Davis, R. W. (1980). Construction of a genetic linkage map in man using restriction fragment length polymorphisms. *American Journal of Human Genetics*, **32**, 314–331.

Komatsu, Y. (1952a). Probability-theoretic investigations on inheritance. VII_2. Non-paternity problems. *Proceedings of the Japanese Academy*, **28**, 105–108.

Komatsu, Y. (1952b). Probability-theoretic investigations on inheritance. VII_4. Non-paternity problems. *Proceedings of the Japanese Academy*, **28**, 112–115.

Neel, J. V., Satoh, C., Goriki, K., Fujita, M., Takahashi, N., Asakawa, J. & Hazama, R. (1986). The rate with which spontaneous mutation alters the electrophoretic mobility of polypeptides. *Proceedings of the National Academy of Sciences USA*, **83**, 389–393.

Skolnick, M. (1986). Cited in Ott, J. (1986). A short guide to linkage analysis. In *Human genetic diseases*, ed. K. E. Davies, pp. 20–21. Oxford: IRL Press.

Yasuda, N. (1978a). The sampling variance of the linkage disequilibrium parameter in multi-allele loci. *Heredity*, **42**, 155–163.

Yasuda, N. (1978b). Estimation of haplotype frequency and linkage disequilibrium parameter in the HLA system. *Tissue Antigens*, **12**, 315–322.

Yasua, N. (1984). A note on gene frequency estimation in the ABO and ABO-like system. *Japanese Journal of Human Genetics*, **29**, 371–380.

Yasuda, N. & Kimura, M. (1968). A gene counting method of maximum likelihood for estimating gene frequences in ABO and ABO-like systems. *Annals of Human Genetics*, **31**, 409–420.

Yasuda, N. & Tsuji, K. (1975). A counting method of maximum likelihood for estimating haplotype frequency in the HLA-systems. *Japanese Journal of Human Genetics*, **20**, 1–15.

12 *Genetic affinities of human populations*

NARUYA SAITOU, KATSUSHI TOKUNAGA AND KEIICHI OMOTO

Introduction

Differentiation of human populations does not necessarily follow the predictions from a simple model of population fission, because gene flow between them may occur after relatively long isolation. This situation is quite different from the phylogenetic tree of different species, where no gene migration is assumed after speciation. Thus for describing the genetic relationship of populations, instead of a dendrogram (rooted tree), which would describe the phylogenetic tree of populations under the assumption of no migration after fission, a network (unrooted tree) of genetic affinity seems to be more appropriate (Figure 12.1).

The neighbour-joining (NJ) method (Saitou & Nei, 1987), in which the principle of minimum evolution is used, may be suitable for constructing genetic affinity networks of populations. The NJ method does not assume constancy of the evolutionary rate, and it has been shown by computer simulation that it is efficient in

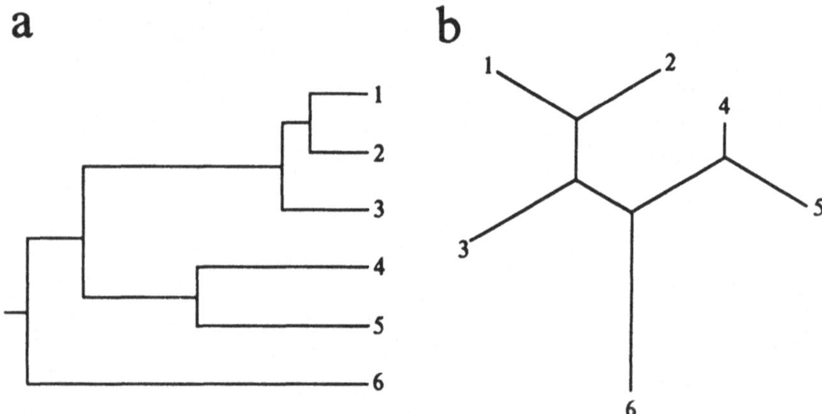

Figure 12.1. A rooted tree (a) and an unrooted tree (b) of 6 populations.

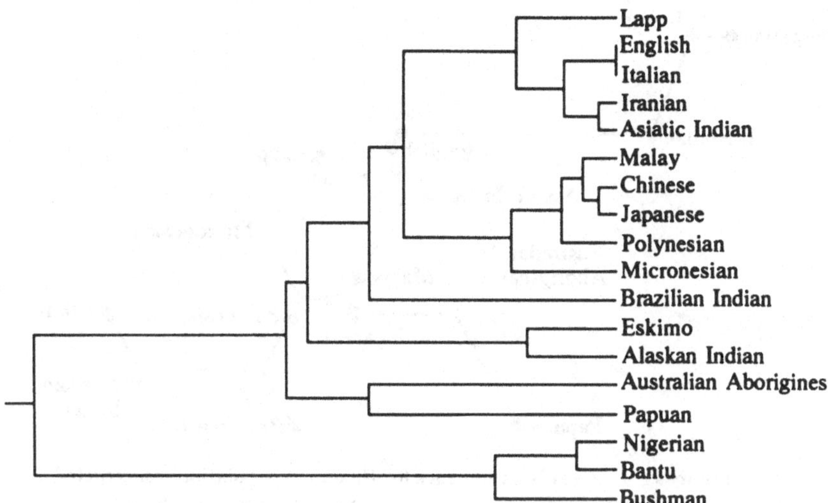

Figure 12.2. A dendrogram for 18 human populations constructed using UPGMA (modified from Nei & Roychoudhury, 1982).

reconstructing the true phylogenetic trees (Saitou & Nei, 1987; Sourdis & Nei, 1988; Saitou & Imanishi, 1989). While in these studies the evolution of nucleotide sequences was simulated, it is likely that the NJ method is also efficient in reconstructing trees from genetic distance matrices based on allele frequency data. In the present study this method is applied to three sets of genetic distance data and the resulting affinity networks are compared with those obtained by other tree-making methods.

Genetic distance data on 18 human populations

Nei and Roychoudhury (1982) compiled allele frequency data on world populations and computed Nei's (1972) genetic distances among 18 human populations based on 23 genetic loci (their Table XI). They constructed a dendrogram of these populations (Figure 12.2) by using UPGMA (Sokal & Sneath, 1963). In this dendrogram, three subSaharan African populations (Nigerian, Bantu and Bushman) stand apart from the remaining populations, and Caucasoid (Lapp, English, Italian, Iranian and Asiatic Indian) and Asian Mongoloid populations (Malay, Chinese, Japanese, Polynesian and Micronesian) each constitute a monophyletic group. However, Amerind populations (Eskimo, Alaskan Indian and Brazilian Indian), who are considered to be genetically close to Asian Mongoloids, are located outside the Caucasoid-Asian Mongoloid cluster. Nei and Roychoudhury (1982) attributed this

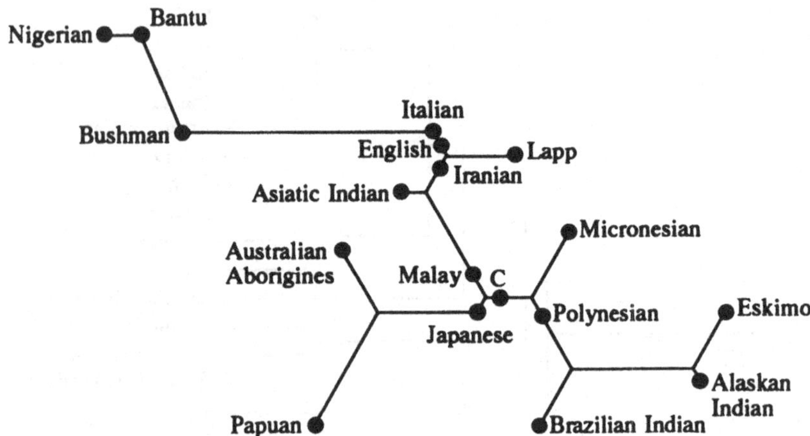

Figure 12.3. An affinity network for 18 human populations constructed using the Wagner distance method (modified from Nei & Saitou, 1986). C = Chinese.

unusual Amerind location to the inbreeding of Amerind populations. Australoid populations (Aborigines and Papuan) were also shown to be genetically similar to Asian Mongoloids (Omoto, 1982). An essentially similar dendrogram was obtained when allele frequency data for the HLA-A and B loci were added (Ryman *et al.*, 1983).

Nei and Saitou (1986) applied the Wagner distance method (Farris, 1972) to the genetic distance matrix of Nei and Roychoudhury (1982). The affinity network they obtained is modified in Figure 12.3, in that negative branch lengths are converted to positive ones as follows: branches of negative length are omitted, and those with positive lengths are drawn proportional to their lengths (=genetic distances). This rule also applies to the following figures. There are marked differences in clustering of populations between the dendrogram (Figure 12.2) and the network (Figure 12.3). Although the three African populations remain monophyletic, neither Caucasoid nor Asian Mongoloid populations are monophyletic. On the other hand, Amerind populations are now monophyletic, and this cluster as well as the Australoid cluster are closer to Asian Mongoloid than to Caucasoid, which is located between African and the other populations. In the following, Asian Mongoloid, Amerind and Australoid as a whole will be called 'Pan-Mongoloid'.

The NJ method was applied to the genetic distance matrix of Nei and Roychoudhury (1982), and the network obtained is shown

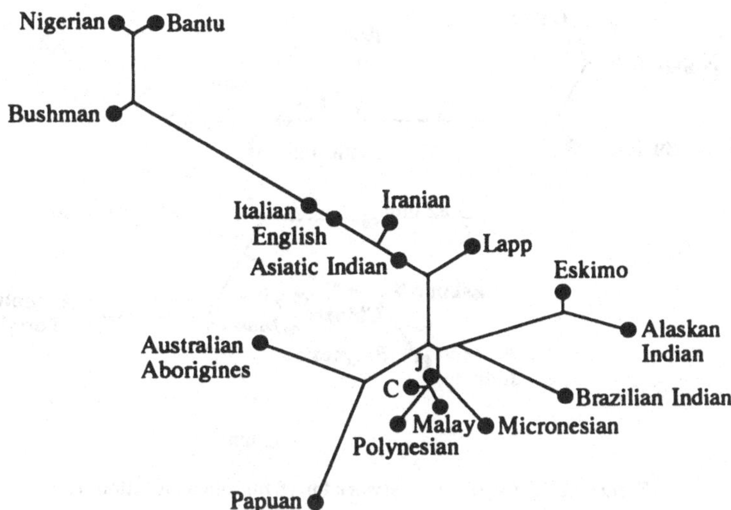

Figure 12.4. An affinity network for 18 human populations constructed using the NJ method (distance matrix data from Nei & Roychoudhury, 1982). C = Chinese, J = Japanese.

in Figure 12.4. In its general features the clustering pattern is similar to that of Figure 12.3; African populations are monophyletic and are far from the remainder, Caucasoid populations are located between African and Pan-Mongoloid, and Amerind populations are monophyletic. Because both the Wagner distance and the neighbour-joining (NJ) methods are intended to produce minimal evolution networks, these similarities are expected. However, there are some important differences. Asian Mongoloid populations are tightly clustered to become a monophyletic group in the NJ network, and the Lapp, instead of the Asiatic Indian, is located between the other Caucasoid populations and the Pan-Mongoloid cluster. Some branch lengths also show differences between Figures 12.3 and 12.4.

Li's (1981) method and modified Farris method (Tateno *et al.*, 1982) were also applied to the same distance matrix data (Figures 12.5 and 12.6, respectively). A UPGMA dendrogram was first constructed in Li's (1981) method, and the dendrogram was then modified by distance transformation. Therefore, the position of the root of Figure 12.5 is identical with that of the UPGMA dendrogram, but Figure 12.5 was drawn as if there were no root. Tateno *et al.*'s (1982) method is a modification of Farris' (1972) Wagner distance method. Although these two trees (Figures 12.5 and 12.6) seem generally to resemble Figures 12.3 and 12.4, there

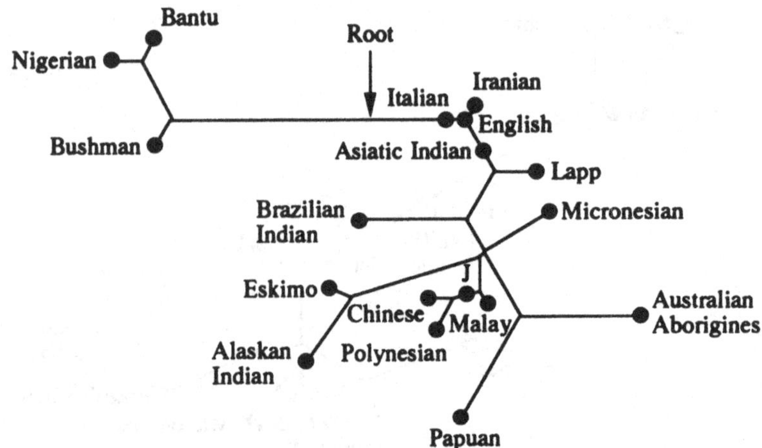

Figure 12.5. An affinity network for 18 human populations constructed using Li's method (distance matrix data from Nei & Roychoudhury, 1982). J = Japanese.

are some conspicuous differences in the Pan-Mongoloid cluster, especially in the location of the Brazilian Indian population. This is located between Caucasoid and the remaining Pan-Mongoloid populations in Figure 12.5, whereas it becomes a part of the Asian Mongoloid cluster in Figure 12.6. Monophyletic grouping of Amerind poulations (Brazilian Indian, Eskimo, and Alaskan Indian) observed in Figures 12.3 and 12.4 seems to be more reasonable if the geographical distribution of these populations is considered.

Some of the clustering features are shared by all of the five figures (one dendrogram, Figure 12.2, and four networks, Figures 12.3-6) produced from the same genetic distance matrix. These are the African cluster (Bantu, Bushman, and Nigerian), the Australoid cluster (Australian Aborigines and Papuan), and the North Amerind cluster (Eskimo and Alaskan Indian). These concordant clusterings seem reasonable in view of the geographical distribution of the populations. Because the amount of migration between a pair of populations is expected to be inversely related to the geographical distance between them, the geographical proximity of populations seems to be a good indicator of their genetic affinity.

If the geographical proximity of populations is used as a criterion for comparing different networks, the Amerind cluster in Figure 12.3 (Wagner distance network) and Figure 12.4 (NJ network), and the Asian-Mongoloid cluster observed in Figure 12.2

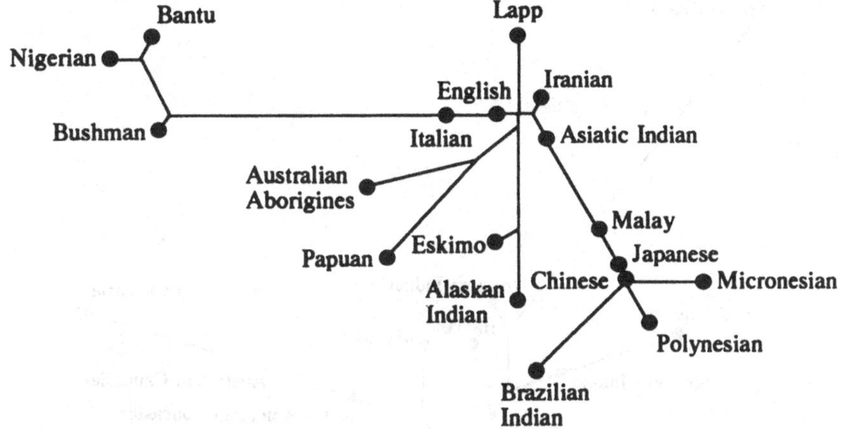

Figure 12.6. An affinity network for 18 human populations constructed using modified Farris method (distance matrix data from Nei & Roychoudhury, 1982).

(UPGMA dendrogram) and Figure 12.4, may represent the true picture. The Caucasoid populations constitute a monophyletic cluster in the UPGMA dendrograms, and they are located between the African cluster and the Pan-Mongoloid cluster in all four networks (Figures 12.3-6). This suggests that Caucasoid populations have experienced gene migrations from Negroid and Mongoloid populations. This possibility was also suggested by Nei and Livshits (1989) who studied the relationship between three major groups of humans.

Genetic distance data on 30 populations from HLA loci

The second set of data analysed is taken from Wakisaka *et al.* (1986) who estimated Nei's (1972) genetic distances for 30 populations based on the HLA allele frequencies, using at locus A 16 antigens, B 32 antigens, C 6 antigens, DR 10 antigens, DRw52/53 2 antigens, and DQ 3 antigens. Wakisaka *et al.* (1986) constructed a UPGMA dendrogram and a modified Farris network based on the distance matrix. A network constructed using the NJ method from the same distance matrix is shown in Figure 12.7. As in the case of the previous analysis, there are marked differences between the UPGMA dendrogram and the two networks. For example, the American Black population is known to be derived from admixture between African Black and European; its intermediate position is realised in the two networks

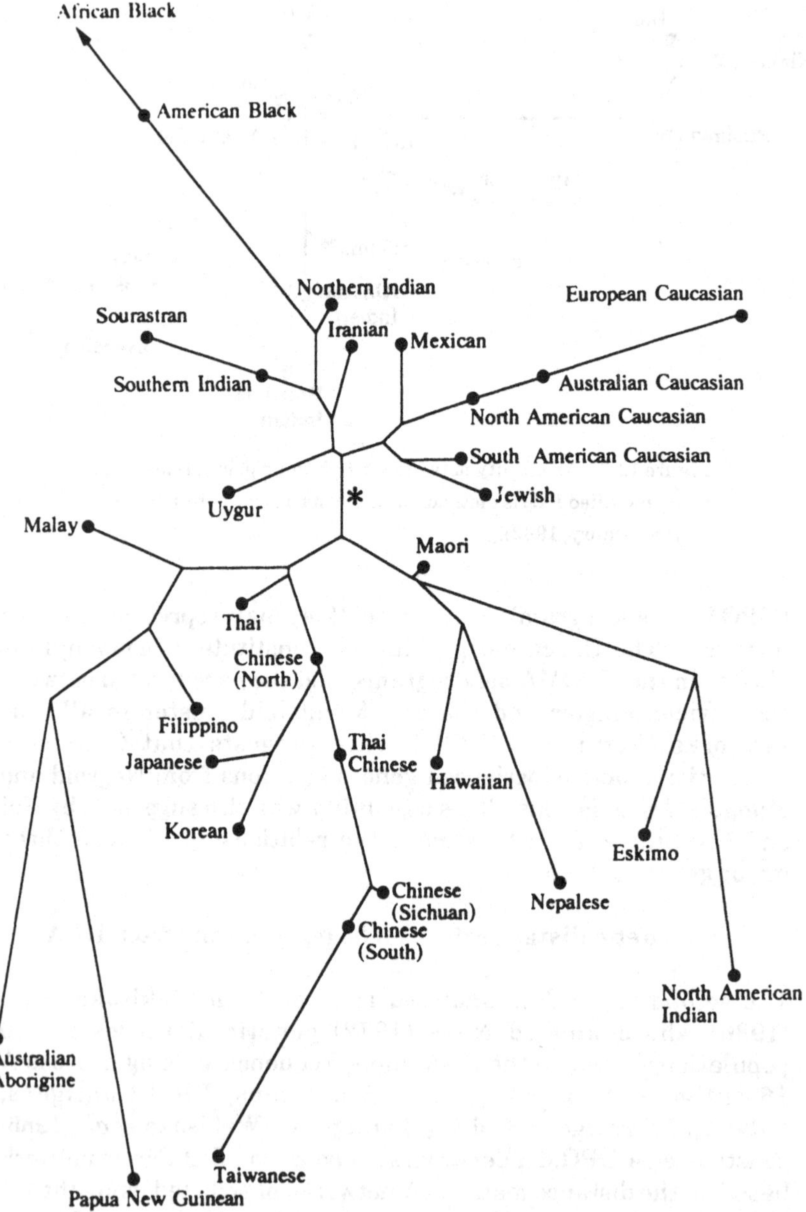

Figure 12.7. An affinity network for 30 populations constructed using the NJ method (distance matrix data from Wakisaka *et al.*, 1986).

(e.g. Figure 12.7), while it clusters with Caucasian populations in the UPGMA dendrogram (Figure 2 of Wakisaka *et al.*, 1986). The position of the Amerind cluster (Eskimo and North American Indian) is also problematic; they are outside the cluster containing Caucasian, Asian Mongoloids, and Polynesian. The situation is similar to the UPGMA dendrogram (Figure 12.2) based on the 23 genetic loci, as discussed above.

Although the two networks are more similar to each other than to the UPGMA dendrogram, there are many differences between the network constructed by the modified Farris method and that constructed by the NJ method. In the NJ network of Figure 12.7, Pan-Mongoloid populations are monophyletic, being separated from the remaining populations by the internal branch denoted by an asterisk. On the other hand, Caucasian populations are located between the Pan-Mongoloid and the African clusters. This is consistent with the NJ network of Figure 12.4 for a different set of data. A similar analysis for mitochondrial DNA data (Saitou & Harihara, 1992) also shows this tendency. However, the Australoid cluster (Australian Aborigine and Papua New Guinean), which is a subcluster of the Pan-Mongoloid cluster in the NJ network, is outside the Caucasian-Mongoloid cluster in the modified Farris network. The position of Japanese also differs in the two networks. In the NJ network (Figure 12.7), Japanese and Korean group together, as a subcluster of one containing various Chinese populations, while in the modified Farris network the Japanese are located with the Amerind cluster. In summary, the clustering of populations in the NJ network seems to reflect the geographical locations of these populations more closely than those in the modified Farris network or in the UPGMA dendrogram. Interestingly, this conclusion is the same as that of the previous analysis.

There is, however, a noteworthy inconsistency between the NJ network and geographical location in the clustering of populations. Hawaii and Nepal are geographically far apart, but the populations living there form a cluster in the NJ network (Figure 12.7). This genetic proximity is also observed in the modified Farris network, and the Polynesian group containing Hawaiian and Maori clusters with Nepalese in the UPGMA tree (Wakisaka *et al.*, 1986). The genetic distance matrix shows Hawaiian and Malay to be equally close, indeed the closest, to Nepalese. Because these genetic distances were computed from allele frequency data at six HLA loci that are tightly linked on chromosome 6 and not independent of each other, the effect of random genetic drift, or any other differentiating process, may be

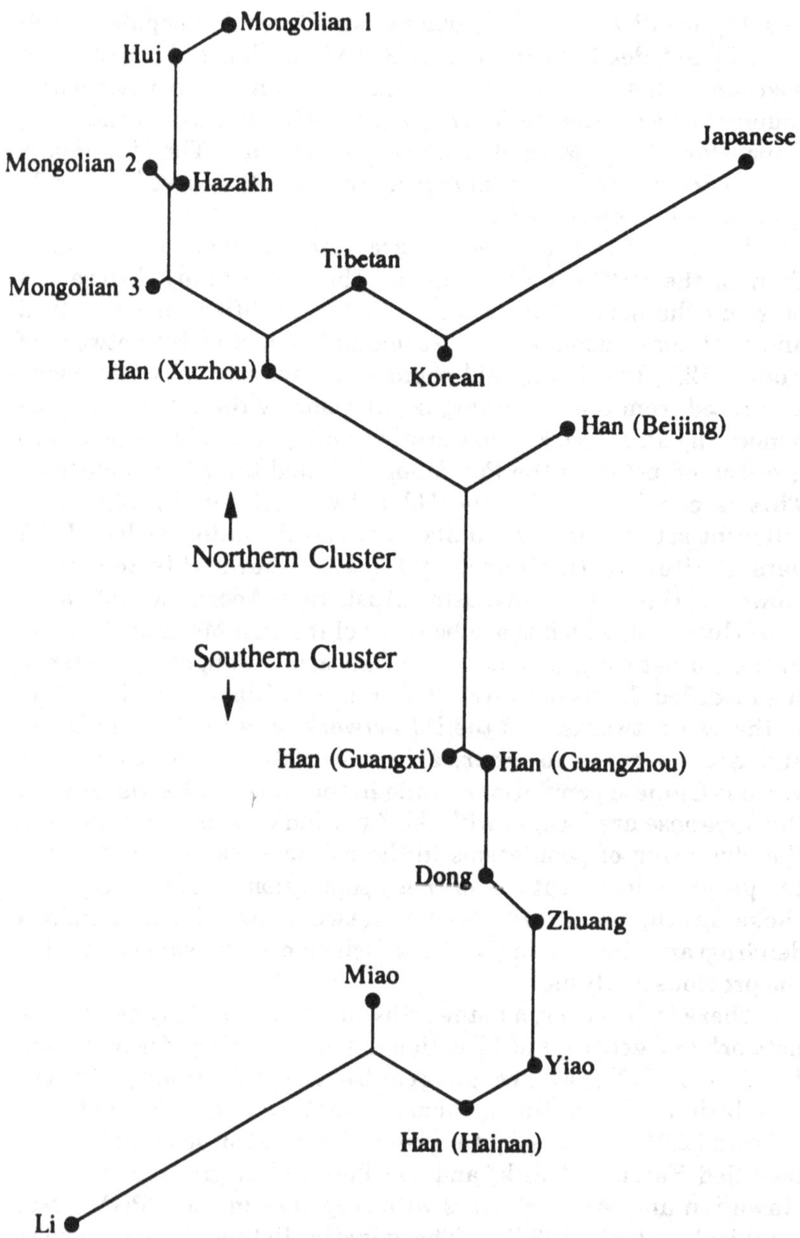

Figure 12.8. An affinity network for 18 east Asian populations constructed by using the NJ method (data from Sun *et al.*, 1986; Aizawa *et al.*, 1986).

exaggerated and apparently larger than that for a set of six unlinked loci. If this is the case, peoples who diverged long ago may show quite similar allele frequencies only by chance. The seemingly close affinity between Hawaiian and Nepalese populations may be so caused.

Genetic distance data on 18 East Asian populations from HLA loci

Allele frequency data were also analysed for the HLA-A, B and C loci of Sun *et al.* (1986) for 16 Chinese populations as well as those of Japanese and Korean. The data for these last two populations were taken from Aizawa *et al.* (1986). Nei's (1972) genetic distances were computed based on the combined allele frequency data (not shown), and an affinity network was constructed using the NJ method (Figure 12.8).

Two clusters can be recognised corresponding to southern and northern populations of East Asia. In the southern cluster, Li and Miao populations are ethnic minorities of Hainan Island, and Dong, Zhuang, and Yiao populations are also ethnic minorities of the southern part of mainland China. Han populations living in the southern provinces of Hainan, Guangxi, and Guangzhou also belong to the southern cluster, while Han of Beijing and Xuzhou belong to the northern cluster. The northern cluster also includes three Mongolian populations, Hui, Hazakh, and Tibetan. Interestingly, the Korean population is closest to the Japanese, and they also belong to the northern cluster. A similar pattern was reported by Omoto *et al.* (1989) who constructed an NJ network based on the allele frequency data of blood group, serum protein and red cell enzyme loci for several Chinese populations and Japanese. Extensive data on Gm haplotype frequencies (Matsumoto, 1987) also show marked differences between northern and southern Chinese populations, and Japanese and Koreans were closer to the northern Chinese populations. These results suggest that significant proportions of the present populations of Japan and Korea are the descendants of migrants from northern Chinese or their related populations. The good correlation between geographical proximity and the genetic affinity of populations is also observed in this set of data.

Conclusion

For describing the genetic relationship of populations, a network (unrooted tree) of genetic affinity seems to be more appropriate than a dendrogram (rooted tree). The neighbour-joining method, in which the principle of minimum evolution is used for

constructing networks, appears suitable for this purpose. This method was applied to three sets of genetic distance data and the resulting affinity networks were compared with those obtained by other methods. It was shown that the genetic affinity networks generally reflect the geographical location, isolation and migration of human populations.

References

Aizawa, M., Natori, T., Wakisaka, A. & Konoeda, Y. (1986). *HLA in Asia-Oceania 1986*. Sapporo: Hokkaido University Press.

Farris, J. S. (1972). Estimating phylogenetic trees from distance matrices. *American Naturalist*, **106**, 645–668.

Li, W. -H. (1981). A simple method for constructing phylogenetic trees from distance matrices. *Proceedings of the National Academy of Sciences*, USA, **78**, 1085–1089.

Matsumoto, H. (1987). Characteristics of the Mongoloid and neighboring populations on the basis of the genetic markers of immunoglobulins (in Japanese). *Journal of the Anthropological Society of Nippon*, **95**, 305–324.

Nei, M. (1972). Genetic distance between populations. *American Naturalist*, **106**, 283–292.

Nei, M. & Livshits, G. (1989). Genetic relationships of Europeans, Asians and Africans and the origin of modern *Homo sapiens. Human Heredity*, **39**, 276–281.

Nei, M. & Roychoudhury, A. K. (1982). Genetic relationship and evolution of human races. *Evolutionary Biology*, **14**, 1–59.

Nei, M. & Saitou, N. (1986). Genetic relationship of human populations and ethnic differences in relation to drugs and food. In *Ethnic differences in reaction to drugs and xenobiotics*, ed. W. Kalow, H. W. Goedde & D. P. Agarwal, pp. 21–37. New York: Alan R. Liss, inc.

Omoto, K. (1982). Allelic diversity of human populations in the Asia-Pacific area. In *Molecular evolution, protein polymorphism, and the neutral theory*, ed. M. Kimura, pp. 193–214. Tokyo: Japan Scientific Societies Press.

Omoto, K., Saitou, N., Misawa, S., Yamazaki, K., Du, R., Zu, J., Zhao, H., Zhang, Z., Wei, X., Niu, K., Hao, L., Du, C. & He, X. (1989). Japanese-Chinese joint study on national minorities in Hainan Island (second study). 1. Blood genetic markers. *Journal of the Anthropological Society of Nippon*, (Abstract), **97**, 292.

Ryman, N., Chakraborty, R. & Nei, M. (1983). Differences in the relative distribution of human gene diversity between electrophoretic and red and white cell antigen loci. *Human Heredity*, **33**, 93–102.

Saitou, N. & Harihara, S. (1992). Gene phylogeny and population phylogeny reconstructed from human mitochondrial DNA data. In *Human evolution in the pacific region*, ed. C. K. Ho, E. Kranz & M. Stoneking. Washington; Washington State University Press.

Saitou, N. & Imanishi, T. (1989). Relative efficiences of the Fitch-Margoliash, maximum-parsimony, maximum-likelihood, minimum-evolution, and neighbor-joining methods of phylogenetic tree construction in obtaining the correct tree. *Molecular Biological Evolution*, **6**, 514–525.

Saitou, N. & Nei, M. (1987). The neighbor-joining method: a new method for reconstructing phylogenetic trees. *Molecular Biological Evolution*, **4**, 406–425.

Sokal, R. & Sneath, P. H. P. (1963). *Principles of numerical taxonomy*. San Francisco: W -H Freeman, Co.

Sourdis, J. & Nei, M. (1988). Relative efficiencies of the maximum parsimony and distance-matrix methods in obtaining the correct phylogenetic tree. *Molecular Biological Evolution*, **5**, 298–311.

Sun, Y., Lee, J., Gao, X., An, J., Li, S., Song, C., Shi, Q. & Li, Z. (1986). HLA antigens in Chinese populations. In *HLA in Asia-Oceania 1986*, ed. M. Aizawa, T. Natori, A. Wakisaka & Y. Konoeda, pp. 502–510. Sapporo: Hokkaido University Press.

Tateno, Y., Nei, M. & Tajima, F. (1982). Accuracy of estimated phylogenetic trees from molecular data. I. Distantly related species. *Journal of Molecular Evolution*, **18**, 387–404.

Wakisaka, A., Hawkin, S., Konoeda, Y., Takada, A. & Aizawa, M. (1986). Anthropological study using HLA antigen frequencies as a genetic marker. In *HLA in Asia-Oceania 1986* ed. M. Aizawa, T. Natori, W. Wakisoka & Konoeda, pp. 197–208. Sapporo: Hokkaido University Press.

13 *Inherited neurological diseases in island isolates in southern Japan*

KIYOTARO KONDO

Introduction

Not long ago, in many areas in rural Japan there lived almost closed endogamous populations engaged in primary industries. These were local isolates, which barely exist as such today. Yet in particular areas, there are still sometimes unique inherited diseases, and the frequency even of less unusual diseases is sometimes grossly elevated, deriving from the disease pattern of the past when these areas were the home of small or isolated populations. One such area is the Amami islands, inhabited by small and isolated populations characterised by unusual patterns of inherited neurological diseases, especially Ryukyu spinal progressive muscular atrophy which is known only in this area among the descendants of the ancient local rulers.

The Amami islands and their people

An arc of islands, known as the south-western islands, links Kyushu, Japan and the island of Formosa (Taiwan). Administratively the northern half of these islands, including the Amami group, belongs to Kagoshima Prefecture, while the others comprise Okinawa Prefecture, formerly known as the Ryukyus (Figure 13.1).

The neolithic Jomon hunting-gathering culture covered almost the whole of Japan, including most of these islands but excluding northern Hokkaido. The Jomon people depended on the products of the deciduous forests for their staple foods. Following the immigration of newcomers from south-east Asia bringing rice farming, there arose the Yayoi Bronze Age rice-farming culture, and from mixture between the immigrants and the Jomon people sprang the present day Japanese. But people in the south-western islands as well as in northern Japan remained relatively unmixed and retained their hunting-gathering economy because these areas were not suitable for the crude methods of rice farming then

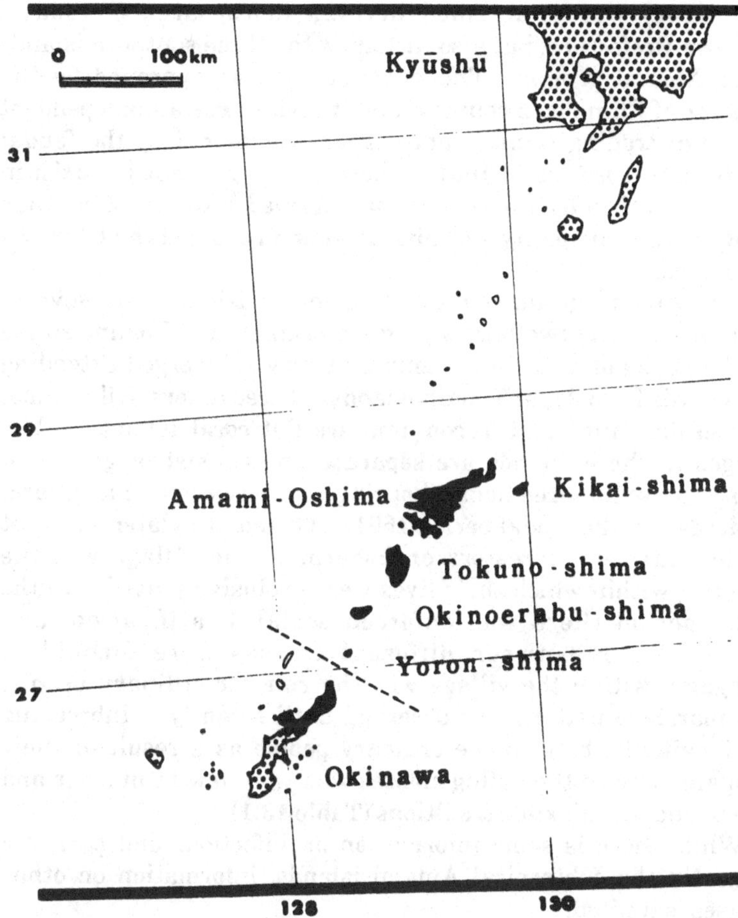

Figure 13.1. The area of survey is shown black. Islands north of the Amami islands are part of Kagoshima Prefecture, Okinawa island and other southerly islands not shown make up Okinawa Prefecture.

available. The patterns of restriction fragment length polymorphisms in various peptides suggests genetic proximity between the inhabitants of both extremities of the Japanese archipelago, namely in the south the people in Okinawa Prefecture and in the north the Ainu, the aborigines of Hokkaido island.

Though Japan has existed as a unified country since about the 6th century AD, neither the south-western islands nor Hokkaido were effectively controlled by the central administration in the early centuries. The inhabitants of the southern islands remained savages until, after many inter-tribal wars, the Ryukyu principality was established in 1429 in the area of today's Okinawa Prefecture, but also including the three southern islands of the Amami group. The principality was approved by the Emperor of China as a country, and indeed it was an independent successful trading country until it was conquered by the feudal lord from Kagoshima in 1601. Thereafter it continued to exist in name until officially annexed by the reformed Japan in 1879, when the Ryukyu principality was abolished and renamed the Okinawa Prefecture.

The Amami group contains five major islands and several small ones. The two largest, Amami-oshima and Tokuno-shima are the peaks of a range of mountains now submerged extending southwards from Japan's main islands. Three others, Kikai-jima, Okinoerabu-shima and Yoron-jima are flat coral islands. The villages in these islands are separate and consist of groups of twenty to two hundred homes housing populations of up to several hundreds. In the feudal period 1601-1868 and even later, for most people - farmers, foresters or fishermen, the village was the universe, within which their lives were exclusively lived. In the feudal period there was enforced social stratification, and intermarriages between different classes were forbidden. Endogamy within the village was the rule for ordinary people. This marriage pattern was observed until recently. Inbreeding was inevitable, both in the ordinary people as a result of their endogamy and in the ruling classes who were few in number and were bound by inflexible traditions (Table 13.1).

While there is some information on infectious and parasitic disease in the subtropical Amami islands, information on other diseases is limited.

Some neurological diseases in southern Japan

There have been a number of reports from southern Kyushu and especially from Kagoshima and Okinawa Prefectures. All neurological diseases except headache, neuralgia and other purely subjective complaints, were listed in a population survey carried out in the town of Setouchi, Amami-oshima in April 1962. Among 7,101 males and 8,299 females, there were respectively 104 (1.44%) and 98 (1.18%) cases with objectively recognised neurological diseases. The prevalences of selected non-genetic diseases were comparable with those found elsewhere (Table 13.2), despite grossly different climatic and socioeconomic conditions.

HTLV-1 associated myelopathy, abbreviated to HAM, is a recently recognised disease entity caused by prolonged infection with human T-cell lymphotrophic virus 1 (HTLV-1). It has an

from onset at 40 to 50 years, nerve biopsy showing axonal degeneration.

d.* Hereditary motor neuropathy, Amami-oshima type. This is a progressive distal myopathy, of which ten cases were found in Amami-oshima. These included a pair of sisters, but otherwise the cases were sporadic. Onset is juvenile, progression is slow, and most patients lose the ability to walk by about 30 years of age (Tsubaki *et al.*, 1964).

e.* Sankwa (small foot) disease. Large families with this dominant condition were found in Kikai-jima, with a total of 61 cases. Onset is in infancy, with distal muscle atrophy that progresses slowly. Even in the seventh decade patients are able to walk using a cane. The disease is characterised by *pes planus* (Osame & Nakashima, 1986).

f.* Ryukyu muscular atrophy (McKusick catalogue number 27120). A special study was made of this.

Ryukyu progressive spinal muscular atrophy

This is a variant of progressive spinal muscular atrophy, and is known in only two islands, Tokuno-shima and Okinoerabu-shima, in the Amami group (Kondo *et al.*, 1970). A field survey of neuromuscular diseases was made which eventually disclosed this disorder.

Method

The area of survey consisted of all the Amami islands and the northern part of Okinawa island together with the neighbouring islets. The survey covered all those (287,727) who were alive on the prevalence day (1st October, 1965) in the area. The target diseases were all those neuromuscular disorders which cause progressive muscular atrophies. Cases were found by the successive information method as follows (Tsubaki *et al.*, 1963).

a. Provisional cases were found from the local records for the disabled, and for handicapped school children, and from local practitioners.

b. These lists were reviewed with the chief or health officer of each settlement, to whom the target diseases were described, and additional possible cases thus collected.

c. The homes of all suspected cases were visited, the patients were re-examined, or where they were not available their relatives were asked about their condition.

d. At each home questions were asked about other cases with similar diseases.

e. A final visit ensured that no new provisional cases had been added.

This method was considered to be particularly suited for the survey. The settlements are closely knit, endogamous communities, and those who have a disability such as muscular atrophy are well known to the inhabitants. Later a door-to-door census was conducted of all the inhabitants of five settlements chosen at random from the 56 settlements in Amami-oshima, and the results were compared with those from the successive information method. The results of the two methods proved in almost complete agreement.

Clinical results

Out of the 59 cases of neuromuscular disease identified in the field survey, 12 cases were Duchenne and one facio-scapulo-humeral muscular dystrophy, 2 were polymyositis, 10 were Charcot-Marie-Tooth disease, and 2 were amyotrophic lateral sclerosis. In one case in northern Okinawa, diagnosis was not established. In 31 cases there was an obscure disease that was different from any other known neuromuscular disease. It was later called the Ryukyu type of progressive spinal muscular atrophy, not only on account of its unique clinical features but on account of the pedigree of the family derived from the rulers of the Ryukyu principality.

The cases were numbered serially, cases 1-23 being found in Tokuno-shima and cases 24-31 in Okinoerabu-shima. Of the 31 cases, 22 were male and 9 female. The age of the patients at the time of the survey ranged from 2-56 years, and their average age was 32.5. The onset was early in life, with flaccid bilateral weakness of the lower extremities observed soon after birth. The patients walked at 1-8 years of age, but from the very beginning most pressed on the knees with the hands when rising. The gait was waddling. At 4 or 5 years, muscle wasting began in both thighs and developed steadily. At 8-15, wasting of the proximal muscles of the arms was also observed. Progression became very slow when the patients reached middle age. In 8 cases the muscles showed fasciculations. Muscle atrophy was symmetrical, more severe in the lower than the upper extremities, and also proximal rather than distal. In 23 cases winging of the scapulae was observed. The muscles of the head and neck were normal. Tendon reflexes were diminished or lost and no pathological reflexes were elicited. Twenty cases showed *pes cavus* of variable degree.

In laboratory examinations, the serum creatine phosphokinase activity was moderately elevated in all seven cases examined.

Electromyography gave a mixed pattern with giant spike-like discharges with duration of 8-10 ms and amplitude of 3-4 mV. Their form was sometimes polyphasic. Although these patterns were compatible with an anterior horn cell lesion, several records showed short durations and lower amplitudes more suggestive of a myopathic lesion. In the five cases where muscle biopsy was done, the fascicular architecture shown in the transverse sections was somewhat disturbed, and a slight lipomatosis was observed both peri- and endo-mysially. The diameter of the muscle fibres was variable, and discrete bundles of small fibres were seen in some areas. Various types of degenerating muscle fibres were scattered throughout all the sections. Histochemically, the phosphorylase activity was weak in small fibres and the aldolase was generally decreased, while the activity of succinic dehydrogenase was, on the whole, well-maintained. No end-plates were found in the sections stained for cholinesterase. In one sample electron microscopy was possible, which showed that the diameter of the myofibrils was generally reduced and variable. However the structure of the myofilament systems and Z bands was normal. The sarcolemma and nuclear membrane were invaginated in some places, and pinocytosis was observed. Aggregating mitochondria were observed subsarcolemmally, and in the sarcoplasma around them glycogen granules were seen. These patterns indicated long-standing courses of gradual denervation due to anterior horn lesions of the spinal cord, with considerable compensatory reinnervation.

Genetic results

Genetic studies were made, the patients and their families being interviewed and pedigrees obtained. These were checked against the koseki records, the legal family register kept in the local municipal offices. In the genetic analysis, the segregation ratio in the sibships of the patients was tested against that expected with recessive inheritance, and the departure from the expected 1 in 4 ratio was not significant. Among the 25 parental pairs, there were 3 first cousin matings, 2 first cousin once removed, and 1 second cousin mating. The rate of consanguinity, defined as a second cousin mating or closer, is 24%. Advantage was also taken of the Keizu, a special genealogical record maintained since feudal times (1601-1868) intended to maintain the status of the family by clarifying its origin, and a few such records are kept by the descendants even today. Fortunately three Keizu were found relating to the affected families. It was noticeable that there were common ancestors of the E and Y

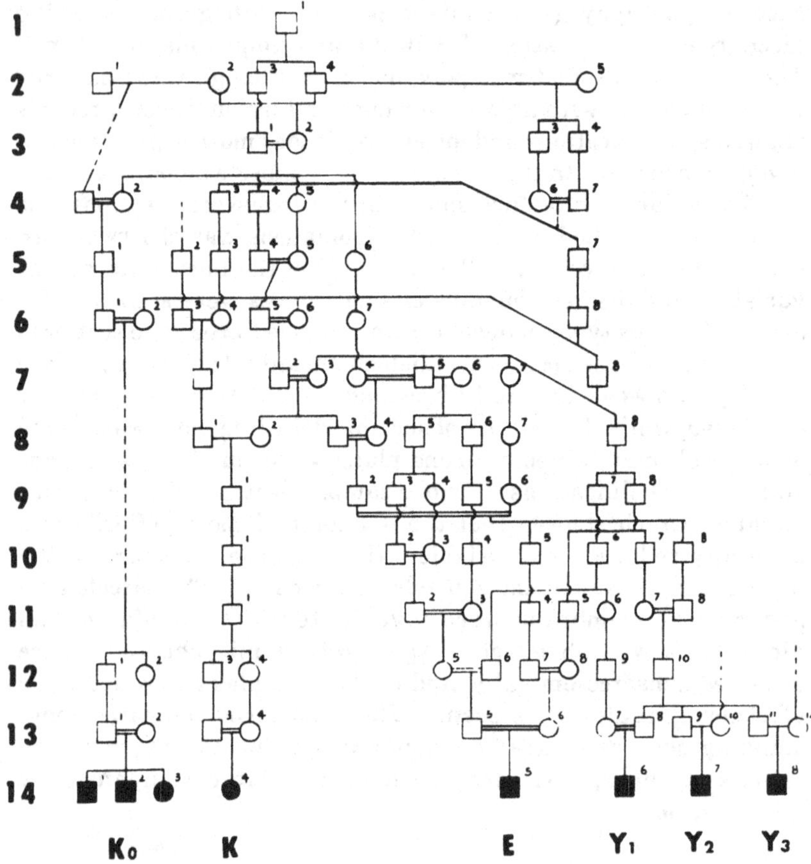

Figure 13.2. An abridged pedigree of eight cases of the Ryukyu spinal progressive muscular atrophy found in Tokuno-shima. 1-1 was Shuri-no-Shu, the first lord of two islands in the early 17th century. Seemingly separate families, the Ko, K, E, Y1, Y2 and Y3 were identified as his direct descendants.

families. Figure 13.2 was drawn using the data from the 3 Keizu together with the additional information in the koseki records. It was recorded in the old local documents that the founder of the Ka family was individual 6.1 in the figure. Therefore out of the 23 cases found in Tokunoshima, at least 8 had as a common ancestor individual 1.1.

Figures 13.3 and 13.4 show the distribution of cases in Tokunoshima and Okinoerabu-shima. The case prevalence (the number of living cases per hundred thousand population) was 47.3 in the former and 31.9 in the latter, phenomenally high for a

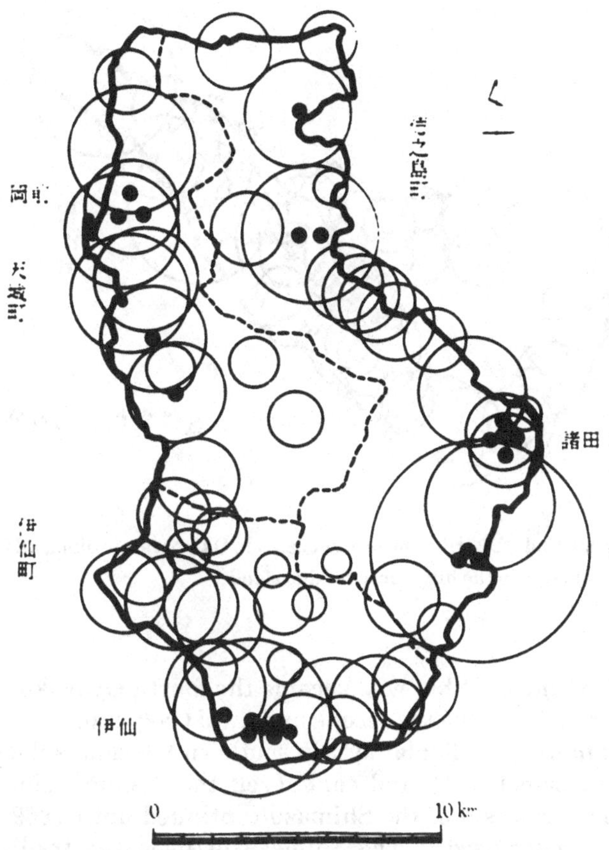

Figure 13.3. Distribution of twenty-three cases in Tokuno-shima; each small black dot represents a case and the diameters of the circles correspond to the square root of the total population of each settlement.

neuromuscular disease. In Tokuno-shima, the distribution was uneven; most cases were found in Shoda and Isen which are small communities. In Okinoerabu-shima all the 8 cases occurred in small communities in the central hilly part of the island. Such distributions were reasonably compatible with the residence and genealogical facts recorded in the Keizu over about 400 years. There was one ancestor (1.1) common to at least 6 affected sibships in Tokuno-shima. His name was Shurino-Shu who was born in the Ryukyus and was appointed as the first Oyayaku, or deputy lord of the island, when Tokuno-shima came under the rule of the Ryukyu principality in the 16th century. He lived in the settlement of Waseura, close to today's Shoda. His wife was a woman from Tokuno-shima. In the next generation, the name of

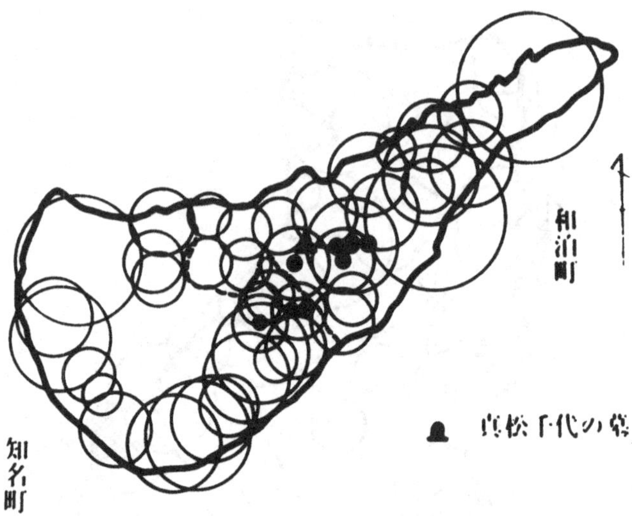

Figure 13.4. Distribution of eight cases in Okinoerabu-shima, all in small settlements in the hilly centre of the island.

2.1 was also Shurino-Shu, who became the fourth Oyayaku. Two other sons of 1.1 became the second and third Oyayaku.

The Shimazu, the feudal lord in southern Kyushu, subjugated the Ryukyu principality and conquered the Amami islands in 1609. The despotism of the Shimazu continued until 1869 when Japan was modernised. The Shimazu utilised the traditional ruling class of the islands to govern them. The Oyayaku lineage therefore held important offices, and because of the enforced social stratification, their families tended to choose their spouses from within the same lineage. Shirama (1659-1737), shown as 6.8 in Figure 13.2 asked the Shimazu authority to move to Isen to explore the area. The family did so, prospered there, and gave rise to the Y and K families.

The distribution of the cases today corresponds well to the activities of these ancestors.

Further observations

Since the report of the disease in 1970, about 10 new local cases have been reported in subsequent surveys in both islands. Of these 2 were known to be related to the 31 cases (Nakahara unpublished). In addition, Nakazato *et al.* (1977) came across a case in Tokyo, a 10 year old boy with a typical Ryukyu progressive spinal muscular atrophy. His parents were from Isen, Tokuno-shima. The author was able to identify his ancestry, and it was established that one of his mother's cousins was Y in Figure 13.2.

Discussion

The Ryukyu progressive spinal muscular atrophy is probably due to a single mutation that occurred at least several hundred years ago, and the dissemination of the gene was restricted by the isolation of the islands. The mild nonfatal disablement and the absence of severe disability in early life means that the patients reproduce before the deleterious effects of the gene become manifest, so that selection against the gene is incomplete, and allows it to persist in the community not only in the heterozygous but also in the homozygous state.

Today there is no longer any separation of the offspring of the lord from other islanders, and consanguineous marriages and marriages within the local community have drastically decreased. Since the gene is recessive in effect, the incidence of the disease will diminish as consanguineous unions diminish. The number of patients born in Tokuno-shima in the twenty years 1915-1934 was 12, but those born between 1935 and 1954 numbered only 6. Emigration of young people however will distribute the gene to other parts of the country, most likely the urban communities of mainland Japan where they will settle, as the case found in Tokyo demonstrates.

The remoteness of these islands, the socioeconomic isolation imposed by their feudal conqueror, give them the status of treasure islands where ancient Japan survives - treasure islands not only for historians but also for genetic study. At least four inherited neurological diseases are unique to this area, including Ryukyu progressive spinal muscular atrophy, for which a lucky chance allowed the origin of the mutation to be traced back to the common ancestors who lived several hundred years ago. Even in an inherited disease that occurs worldwide, it is possible that all patients carry the same gene, copies of the same mutation that occurred independently in various isolates. The evidence seems clear for some disorders that still show remarkable foci of elevated incidence like Huntington's chorea in the New World which appears to have been introduced from some areas in Europe by heterozygous immigrants; this disease is observed elsewhere in the world, usually at very low prevalence rates, without any traceable link with European patients. Yet recent DNA work shows that different mutations may occur within a gene and yet produce the same deleterious effect.

It is probable that man lived in small semi-isolated

communities not only during the long primeval ages but in historical times also, until farming allowed the development of large populations and modern cosmopolitan societies brought panmixis. The pattern of occurrence of Ryukyu progressive spinal muscular atrophy indicates a transitional state, in which a specific mutant gene was confined within an endogamous isolate, but began to spread elsewhere as the isolation broke down.

References

Kondo, K., Tsubaki, T. & Sakamoto, F. (1970). The Ryukyuan muscular atrophy; an obscure heritable neuromuscular disease found in the islands of southern Japan. *Japanese Journal of neurological Science*, **11**, 359–382.

Kurtzke, J. F. (1982). The current neurologic burden of illness and injury in the United States. *Neurology*, **32**, 1207–1214.

Nakazato, H., Kinoshita, M. & Satoyoshi, E. (1977). A case of Ryukyuan muscular atrophy. *Clinical Neurology (Japan)*, **17**, 353–356.

Osame, M. & Furusho, T. (1983). Genetic epidemiology of myotonic dystrophy in Kagoshima and Okinawa districts in Japan. *Clinical Neurology (Japan)*, **23**, 1067–1071.

Osame, M. Matsumoto, M., Usuku, K., Izumo, S., Ijichi, N., Amitani, H., Tara, M. & Igata, A. (1987). Chronic progressive myelopathy associated with elevated antibodies to HLTV-I and adult T-cell leukemia like cells. *Annals of Neurology*, **21**, 117–122.

Osame, M. & Nakashima, H. (1986). Two new forms of familial motor neuron diseases found in southern Japan. *Japanese Journal of Geriatrics (Japan)*, **22**, 28–34.

Tsubaki, T. Kondo, K., Tsukagoshi, H., Takasu, T., Nakanishi, T., Suga, M. & Kuroiwa, Y. (1963). Study of neurologic disorders in Amami-oshima island. *Clinical Neurology (Japan)*, **3**, 394–400.

Tsubaki, T., Kondo, K., Tsukagoshi, H., Takasu, T. & Nakanishi, T. (1964). Study of neurologic disorders in Amami-oshima island (second report). *Clinical Neurology (Japan)*, **4**, 441–447.

14 *Serological and virological evidence for human T-lymphotropic virus type I infection among the isolated Hagahai of Papua New Guinea*

RICHARD YANAGIHARA AND RALPH M. GARRUTO

Introduction

Although human T-lymphotropic virus type I (HTLV-I) was the first human retrovirus to be isolated and the first to be shown to cause a human cancer, adult T-cell leukaemia/lymphoma (ATLL), its importance was overshadowed by the isolation and identification of human immunodeficiency virus as the cause of the acquired immunodeficiency syndrome. A resurgence of interest in HTLV-I, a member of the oncovirinae subfamily of the family Retroviridae, occurred after the fortuitous discovery that patients with endemic tropical spastic paraparesis (TSP) in Martinique had serological evidence of HTLV-I infection (Gessain *et al.*, 1985). Confirmatory data demonstrating IgG antibodies against HTLV-I in sera and cerebrospinal fluids (CSF) of patients with TSP were soon reported from Jamaica (Rodgers-Johnson *et al.*, 1985, 1988), Colombia (Rodgers-Johnson *et al.*, 1985; Zaninovic, 1987), Trinidad (Bartholomew *et al.*, 1986), the Seychelle Islands (Roman *et al.*, 1987) and West Africa (Gessain *et al.*, 1986; Tournier-Lasserve *et al.*, 1987). In addition, patients with a TSP-like disease in southern Japan, designated HTLV-I-associated myelopathy (HAM) (Osame *et al.*, 1987), were found to be infected with HTLV-I. TSP and HAM are now known to represent the same clinical syndrome and are collectively called TSP/HAM, although the term HTLV-I myeloneuropathy seems more appropriate (Rodgers-Johnson *et al.*, 1990).

Since 1983, we have intensified our search for high-prevalence foci of HTLV-I infection, concentrating primarily on isolated populations of the western Pacific. Serological surveys of HTLV-I infection in Melanesia, using screening tests such as enzyme-linked immunosorbent assay (ELISA) and gel particle agglutination assay, indicate moderate to extraordinarily high prevalences of infection in several remote coastal and inland populations having minimal or no contact with Europeans, Japanese or Africans prior to the bleedings (Goudsmit *et al.*, 1987; Kazura *et al.*, 1987; Asher *et al.*, 1988; Brindle *et al.*, 1988; Babona & Nurse, 1988; Hrdy *et al.*, 1989; Currie *et al.*, 1989; Brabin *et al.*, 1989; Armstrong *et al.*, 1990; Garruto *et al.*, 1990). However, the inability to confirm seropositivity by Western analysis in many Melanesian sera (Asher *et al.*, 1988; Hrdy *et al.*, 1989; Currie *et al.*, 1989; Armstrong *et al.*, 1990; Garruto *et al.*, 1990) and the failure of such sera to neutralise a prototype strain of HTLV-I (Asher *et al.*, 1988; Weber *et al.*, 1989) have raised doubts about these reported high seroprevalences. To clarify this issue, we conducted an in-depth investigation among the Hagahai, a remote, hunter-horticulturalist groups in Madang Province of Papua New Guinea (Jenkins, 1988; Jenkins *et al.*, 1989). Our survey, which was conducted among nearly half of the entire Hagahai population, demonstrates that HTLV-I infection, as verified by stringent Western immunoblot criteria (Centers for Disease Control, 1988), is as prevalent among the Hagahai as among individuals living in regions such as southwestern Japan and the Caribbean basin, where ATLL and TSP/HAM are endemic (Yanagihara *et al.*, 1990a). In addition, we have isolated an HTLV-I -related virus from an interleukin 2-dependent T-cell line derived from peripheral blood mononuclear cells of a healthy Hagahai man (Yanagihara *et al.*, 1990b, 1991a). The successful isolation of an HTLV-I-related virus in Melanesia is consistent with our recent identification of a case of HTLV-I myeloneuropathy in a life-long resident of the Solomon Islands (Ajdukiewicz *et al.*, 1989). Intensified searches are under way to document the prevalences of diseases caused by HTLV-I and HTLV-I-like retroviruses in Papua New Guinea and elsewhere in Melanesia (Yanagihara *et al.*, 1991b).

Study population

The Hagahai (their word for 'people'), a 260-member group (1987 census) consisting of five territorial groups or parishes (Aramo or Ginam, Luyaluya, Miyamiya, Mamusi and Pinale), live at altitudes of 350 to 2400m along the northern side of the Yuat River Gorge in the far western corner of the Schrader Range in Madang Province of Papua New Guinea (Figure 14.1). They first made contact with government and missionary workers in December 1983 (Jenkins, 1988; Jenkins *et al.*, 1989). Two neighbouring groups, the Pinai (pop. 281 in 1988) and the Haruai (pop. 750 in

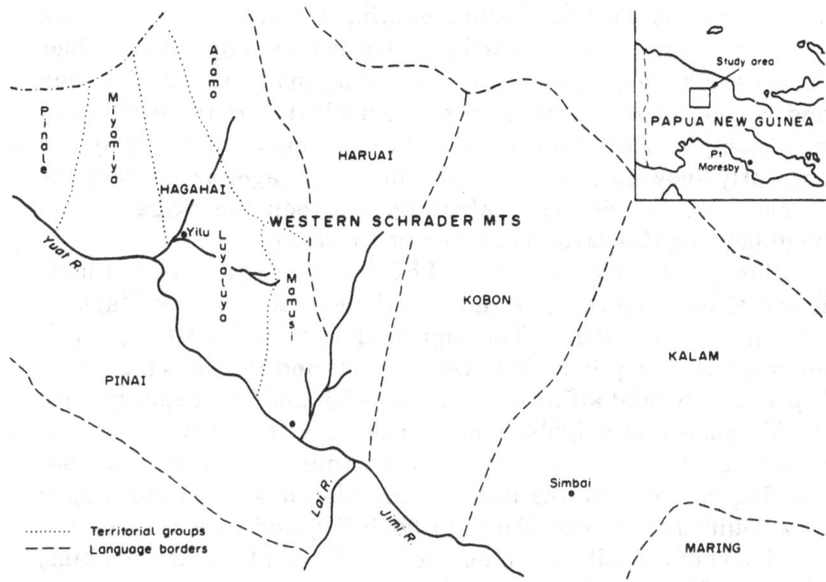

Figure 14.1. Map of the western Schrader Mountains, depicting the locations of the five Hagahai parishes (Pinale, Miyamiya, Luyaluya, Aramo and Mamusi) and their proximity to the neighbouring Haruai and Pinai.

1980), to the south and east respectively, had first contact at least 10 to 15 years previously.

Ethnographic and demographic data, including information on family relationships, were ascertained by interviews conducted initially through tri-lingual Haruai interpreters, and thereafter (in late 1985), in Melanesian pidgin. These data were collected by Dr. Carol L. Jenkins of the Papua New Guinea Institute of Medical Research in Goroka between 1984 and 1988, during a series of field trips, which totalled 12-months of residence among the Hagahai (Jenkins, 1988; Jenkins *et al.*, 1989). Ages, which were estimated based on a variety of information, including physical examination, were most accurate for the younger age groups.

Unlike their more sedentary agrarian neighbours, the Hagahai are primarily hunter-foragers, who practise little horticulture. Their primary crops are banana, taro and yam. They occupy a territory of approximately 750km^2, which includes immense tracts of primary forest and grassland containing abundant wildlife. Population density is less than 1 person per km^2. Hagahai culture is characterised by both highland (house types, clothing, body decoration) and lowland traits (processing of sago, betel-nut chewing). While hunting, the family sleeps in rock shelters or simply constructed lean-tos or on the open forest floor for a week or longer (Jenkins, 1988; Jenkins *et al.*, 1989). Men and women sleep together, often in nuclear family rooms, and monogamy is the rule, although three examples of polygyny are currently known (Jenkins, 1988). Most marriages are elopements or marriage by capture. Marriage between the Hagahai and neighbouring Haruai and Pinai occurs infrequently.

Dermatoglyphic analysis and HLA gene frequencies at the A, B and C loci indicate lowland genetic affinities for the Hagahai (Jenkins *et al.*, 1989). The high frequency of HLA-A11, which approaches nearly 100% in coastal regions and 80% in Sepik areas, support a lowland affinity. Like the highland and Sepik groups, the Hagahai and neighbouring Haruai lack the HLA-A2 antigen associated with recent Austronesian admixture, suggesting that the Hagahai predate the last Austronesian migration into Papua New Guinea, currently dated to 5400 BP, and corroborates the likelihood of a Sepik origin for the Hagahai and Haruai (Jenkins, 1988; Jenkins *et al.*, 1989).

Serological evidence

Enzyme immunoassay

Using a screening ELISA (E.I. Du Pont, Inc., Wilmington, Del) and applying a cut-off equivalent to a ratio of absorbance values between the test serum and the negative control serum of 10 or greater, IgG antibodies against HTLV-I were found in 61 of 120 Hagahai (51%) bled between February 1985 and January 1988. Had the conventional ratio of 5 been used (as recommended by the manufacturer of the ELISA), many more sera would have been considered positive (i.e., 73% rather than 51%), but the overall seroprevalence, as verified by Western analysis, would not have been significantly higher.

Western analysis

As determined by an enhanced HTLV-I Western immunoblot (which incorporates a recombinant polypeptide derived from the gp21 transmembrane protein-encoding region of the HTLV-I *env* gene) (Lillehoj *et al.*, 1990), 16 of the 61 ELISA-positive Hagahai sera were positive; all of these sera reacted to HTLV-I *gag*-encoded proteins p19 and p24 and an *env* gene product (the major envelope glycoprotein gp46 and/or the transmembrane glycoprotein gp21). Of the Western immunoblot-confirmed sera, reactivity was more often found to the *env*-encoded transmembrane protein gp21 than to the major envelope glycoprotein gp46. In fact, only four of the Western immunoblot-confirmed sera contained antibodies to gp46. All four sera were from members of a single household, in which both parents (aged 58 years) and two of five children (two sons aged 18 and 28 years) were infected. Two other sons (aged 26 and 38 years) were seropositive by ELISA, but had indeterminate Western immunoblots (one was reactive to *gag* proteins p19, p28, p32, *tax* protein p38 and *env* protein gp21; and the other to *gag* proteins p19, p26, p28, p32 and p53). A 33-year old daughter was seronegative by ELISA. Sera from all four members with reactivity to gp46 also reacted to gp21, and three of the four also possessed antibodies to the *tax* gene product p38.

Of the 61 ELISA-positive sera, 45 (74%) were indeterminate by Western immunoblot; that is, these sera were reactive just to *gag* gene products (either p19 of p24 alone, or p19 and other *gag* proteins such as p28 and p53), or to p19 and the transmembrane protein gp21. The most common band patterns consisted of reactivity to *gag* gene products p19, p26, p28, p32 and p53 (33%) and to *gag* proteins p19, p24, p26, p28, p32 and p53 (16%). Twenty-eight (62%) of the 45 ELISA-positive, Western immunoblot-indeterminate Hagahai sera reacted to three or more *gag*-encoded proteins. Reactivity to p19 and gp21 occurred in four (9%) of these sera.

Of nine ELISA-negative Hagahai sera tested by Western immunoblot, three were negative (no bands), five were indeterminate (four were reactive with p19 and p28 and one with p24 only), and one, a serum from a 17-year old adolescent male, was positive (reactive with p19, p24 and gp21, but with no other *gag* protein or to *tax* p38). Thus, a total of 17 Hagahai had confirmatory Western blots, giving an overall HTLV-I seroprevalence of 17/120 or 14%.

Age- and sex-specific Western immunoblot data indicated an age-related acquisition of infection. HTLV-I seroprevalence, based on Western blot confirmation, was 0% among children, 9% among adolescents and 21% among adults. Although children as young as one and two years of age were positive by ELISA, none of 10 ELISA-positive children (age range, 1-9 years) had confirmatory Western immunoblots. Three children, however, exhibited reactivity to p19 and p24 and one child, in addition, possessed antibodies to *tax* protein p38. Of the 32 adolescents (10-19 year age group), 16 were ELISA-positive and, of these, seropositivity was verified in only three. By contrast, seropositivity was confirmed in 14 of the 35 ELISA-positive adult men and women (20 years or older).

Clustering of HTLV-I infection

Infection tended to cluster in family groups. Of the 24 households studied, five (21%) had only one seropositive member and four households (17%) had two or more members who were seropositive, as confirmed by Western blot. The remaining 15 (62%) households had no seropositive members. Three (50%) of six households in which one or both parents were seropositive contained seropositive children, while only one (11%) of nine households in which neither parent was infected contained serpositive children. HTLV-I seroprevalence, as verified by Western analysis, was significantly higher in families with one or two seropositive parents (29%) than in families in which neither parent was seropositive (3%) ($\underline{P}<0.01$ by χ^2 analysis).

Virological evidence

Virus isolation

Heparinised blood samples, collected in the field from 24 Hagahai men and women, of whom 7 had confirmatory and 17 had indeterminate HTLV-I Western immunoblots, were rushed to the Papua New Guinea Institute of Medical Research in Goroka, where they were processed in a laboratory in which HTLV-I and other human retroviruses had not previously been present. Following mitogen stimulation for two days, lymphocytes were maintained in medium containing 10% interleukin 2 (Advanced Biotechnologies, Inc., Columbia, Md) for 10 to 15 weeks. Supernatant fluids from three cultures contained low levels of reverse transcriptase activity and rare virus particles resembling HTLV-I were observed by thin-section electron microscopy, beginning at one month, but virus yields were too low for further antigenic characterisation.

On re-isolation attempts using cord blood mononuclear cells, HTLV-I antigens were observed by the indirect immunofluorescent antibody technique in one culture, derived from a 20-year old healthy Hagahai man, beginning at two weeks. However, the percentage of viral antigen-bearing cells remained low (<1%) and the cells, while growing as small clusters, replicated sluggishly for five months. Consequently, as a final resort, the few remaining cell clumps were co-cultivated with newly acquired MOLT 3 cells (CRL 1552, American Type Culture Collection, Rockville, Md), an interleukin 2-independent (non-HTLV-I-infected) T-cell line derived from the peripheral blood of a 19-year old male with acute lymphoblastic leukaemia in relapse. This resulted in the establishment of a rapidly growing cell line, designated PNG-1, which grew as a clumpy suspension but remained dependent on interleukin 2. Using sera from individuals with serologically and virologically verified HTLV-I infection and rabbits experimentally infected with HTLV-I, virus-specific fluorescence was found in 10% of cells at 18 days and in more than 85% at 39 days following co-cultivation with MOLT-3 cells. As evidenced by double-label immunofluoresence, the same cells were stained using the above-mentioned antibodies. Donor cord blood mononuclear cells and MOLT-3 cells were consistently negative in such tests.

Antigenic characterisation

HTLV-I antigen expression in PNG-1 cells was distinct from that of MT-2 cells, an interleukin 2-independent (transformed) T-cell line persistently infected with a prototype strain of HTLV-I, derived from a Japanese patient with adult T-cell leukaemia. Unlike MT-2 cells which exhibited robust fluorescence with monoclonal and polyclonal antibodies to native and synthetic peptides of HTLV-I *gag* proteins p19 and p24, PNG-1 cells exhibited no fluorescence. However, PNG-1 cells, like MT-2 cells, were brightly fluorescent with antibodies against the major envelope glycoprotein gp46 and with sera from patients with serologically or virologically verified HTLV-I spastic myelopathy.

By Western analysis, cell lysates of PNG-1 exhibited virus-specific bands at 15, 19, 46 and 53 kilodaltons, using sera from patients with HTLV-I myeloneuropathy and rabbits experimentally infected with HTLV-I, and antiserum prepared against a synthetic peptide of the C-terminus of gp46. No reactivity was found with sera from HTLV-I-seronegative humans and rabbits.

Gene amplification

HTLV-I genomic sequences were detected in DNA extracted from PNG-1 cells by the polymerase chain reaction (PCR), using oligonucleotide primers specific for *gag*, *env* and *tax* sequences of ATK-1, a prototype strain of HTLV-I.

Significance of the evidence

High seroprevalences of HTLV-I infection in Papua New Guinea have been reported since 1987 (Goudsmit *et al.*, 1987; Kazura *et al.*, 1987; Asher *et al.*, 1988; Brindle *et al.*, 1988; Babona & Nurse, 1988; Hrdy *et al.*, 1989; Currie *et al.*, 1989; Brabin *et al.*, 1989; Armstrong *et al.*, 1990; Garruto *et al.*, 1990). In the neighbouring East Sepik region of Papua New Guinea, an area approximately

300km adjacent to that of the Hagahai, an HTLV-I seroprevalence of 26% was found among the Urat and Urim people (Kazura *et al.*, 1987). Recently, an overall HTLV-I seroprevalence of approximately 14% was reported among 624 non-pregnant women of child-bearing age residing in 17 coastal villages in Madang Province (Brabin *et al.*, 1989). However, Western immunoblots were not performed in the former study, and in the latter, verification of seropositivity was based only on reactivity to *gag* proteins. Thus, our results, demonstrating a high HTLV-I seroprevalence among the inland, recently contacted Hagahai in Madang Province, unequivocally establishes that HTLV-I infection is as prevalent in certain regions of Papua New Guinea as in HTLV-I-endemic regions, such as south-western Japan and the Caribbean basin, eliminating previous doubts about the veracity of the reported high prevalences. The historical remoteness of the Hagahai and other high prevalence groups in Melanesia is consistent with an ancient origin of HTLV-I, and combined with the demonstrated low seroprevalence of HTLV-I infection in Micronesia, argues against the notion that infection was acquired from contact with Japanese or Africans (Garruto *et al.*, 1990).

As in other HTLV-I endemic regions, HTLV-I infection among the Hagahai tended to cluster in families, and was more common with increasing age. Infection in childhood is probably facilitated by the close interpersonal contact typical of communal living in Melanesian societies. Acquisition of infection during infancy most likely occurs via virus-infected lymphocytes in breast milk (Kinoshita *et al.*, 1984; Hino *et al.*, 1985). Although this mode of transmission is probably operative among the Hagahai, we were unable to confirm the high prevalence among children by employing stringent Western immunoblot criteria. Infections in later life probably result through sexual or other intimate contact or through contact with contaminated blood.

The cause of the high frequency of indeterminate Western immunoblots among the Hagahai and other Melanesian groups is unknown. Whether they signify specific antibody reactivity against HTLV-I or HTLV-I-related retroviruses during seroconversion is being investigated. Recent studies indicate that reactivity to p19 and gp21 occurs early, with reactivity to p24 occurring much later, in individuals with transfusion-acquired HTLV I infection. Thus, some of the Hagahai with reactivity to p19 and gp21 (called indeterminate in our study) may actually represent early seroconverters. Serological tests of additional serum samples from the Hagahai should clarify this issue. In addition, studies employing PCR and cell culture techniques specific for retroviruses should establish whether or not healthy Hagahai with indeterminate Western immunoblots are infected with HTLV-I or a related retrovirus. Our isolation of an HTLV-I-related virus from a healthy Hagahai is the first step in elucidating this issue (Yanagihara *et al.*, 1990b 1991a). In addition we have isolated HTLV-I-like viruses from unrelated individuals living in widely separated provinces in the Solomon Islands (Yanagihara *et al.*, 1991c,d). Sequence analysis of these newly isolated HTLV-I strains from Papua New Guinea and the Solomon Islands indicate that they constitute a major variant of HTLV-I, which diverge markedly from contemporary cosmopolitan prototypes of HTLV-I from Japan, the Caribbean, the Americas and Africa (Gessain *et al.*, 1991). These novel sequence variants of HTLV-I are consistent with a proto-Melanesian HTLV-I strain of archaic origin, which evolved independently of contemporary cosmopolitan strains. Molecular genetic studies of future isolates of HTLV-I from Melanesia and elsewhere in Oceania will help understanding of the phylogeny of world dissemination of HTLV-I.

Although TSP/HAM and ATLL have not been reported among the Hagahai, this is not unexpected because of the long latency between acquisition of HTLV-I infection and onset of disease, as well as the low frequency with which these diseases occur in infected individuals. Differences in virulence of the Western Pacific strains of HTLV-I, differences in host genetics or lack of reporting may account for the apparent absence of disease. We have, however, recently identified a case of HTLV-I myeloneuropathy in a Solomon Islander (Ajdukiewicz *et al.*, 1989). This patient, who had no history of blood transfusion or intra-venous drug use, had high titers of IgG antibodies to HTLV-I *gag*-, *tax*- and *env*- encoded proteins in both serum and cerebrospinal fluid, and represents the first confirmed case of HTLV-I-caused disease in Melanesia. Intensified surveillance for ATLL and TSP/HAM is clearly warranted, not only among the Hagahai but among other Melanesian populations in which high HTLV-I seroprevalences have been demonstrated.

References

Ajdukiewicz, A., Yanagihara, R., Garruto, R. M., Gajdusek, D. C. & Alexander, S. S. (1989) HTVL-I myeloneuropathy in the Solomon Islands. *New England Journal of Medicine*, **321**, 615–616.

Armstrong, M. Y. K., Hardy, D. B., Carlson, J. R. & Friedlaender, J. S. (1990) Prevalence of antibodies interactive with HTLV-I antigens in selected Solomon Islands populations. *American Journal of Physical Anthropology*, **81**, 465–470.

Asher, D. M., Goudsmit, J., Pomeroy, K. L., Garruto, R. M., Bakker, M., Ono, S. G.,

Elliott, N., Harris, K., Askins, H., Eldadah, Z., Goldstein, A. D. & Gajdusek, D. C. (1988). Antibodies to HTLV-I in populations of the southwestern Pacific. *Journal of Medical Virology*, **26**, 339–351.

Babona, D. V. & Nurse, G. T. (1988). HTLV-I antibodies in Papua New Guinea. *Lancet* **ii**, 1148.

Bartholomew, C., Cleghorn, F., Charles, W., Ratan, P., Roberts, L., Maharaj, K., Jankey, H., Daisley, H., Hanchard, B. & Blattner, W. (1987). HTLV-I and tropical spastic paraparesis. *Lancet*, **ii**, 99–100.

Brabin, L., Brabin, B. J., Doherty, R. R., Gust, I. D., Alpers, M. P., Fujino, R., Imai, J. & Hinuma, Y. (1989). Patterns of migration indicate sexual transmission of HTLV-I infection in non-pregnant women in Papua New Guinea. *International Journal of Cancer*, **44**, 59–62.

Brindle, R. J., Eglin, R. P., Parsons, A. J., Hill, A. V. S. & Selkon, J. B. (1988). HTLV-I, HIV-I, hepatitis B and hepatitis delta in the Pacific and southeast Asia: a serological survey. *Epidemiology and Infection*, **100**, 153–156.

Centers for Disease Control (1988). Licensure of screening tests for antibody to human T-lymphotropic virus type. I. *MMWR*, **37**, 736–747.

Currie, B., Hinuma, Y., Imai, J., Cumming, S. & Doherty, R. (1989). HTLV-I antibodies in Papua New Guinea. *Lancet*, **i**, 1137.

Garruto, R. M., Slover, M., Yanagihara, R., Mora, C., Alexander, S. S., Asher, D. M., Rodgers-Johnson, P. & Gajdusek, D. C. (1990). High prevalence of human T-lymphotropic virus type I infection in isolated populations of the Western Pacific confirmed by Western immunoblot. *American Journal of Human Biology*, **2**, 439–447.

Gessain, A., Barin, E., Vernant, J. C., Gout, O., Maurs, L., Calender, A. & de The, G. (1985). Antibodies to human T-lymphotropic virus type I in patients with tropical spastic paraparesis. *Lancet*, **ii**, 407–410.

Gessain, A., Francis, H., Sonan, T., Fiordano, C., Akani, F., Piquemal, M., Caudie, C., Malone, G., Essex, M. & de The, G. (1987) HTLV-I and tropical spastic paraparesis in Africa. *Lancet*, **ii**, 698.

Gessain, A., Yanagihara, R., Franchini, G., Garruto, R. M., Jenkins, C. L., Adjukiewicz, A. B., Gallo, R. C. & Gajdusek, D. C. (1991). Highly divergent molecular variants of human T-lymphotropic virus type I from isolated populations in Papua New Guinea and the Solomon Islands. *Proceedings of the National Academy of Science USA*, **88**, 7694–7698.

Goudsmit, J., Asher, D. M., Garruto, D. M., Mora, C., Jenkins, C., Pomeroy, K. L., Askins, H. & Gajdusek, D. C. (1987). Antibodies to HTLV-I in populations of the Western Pacific. In *Abstracts of the 16th Pacific Science Congress, Seoul, Korea,* 73.

Hino, S., Yamaguchi, K., Katamine, S., Sugiyama, H., Amagasaki, T., Kinoshita, K., Yoshida, Y., Doi, H., Tsuji, Y. & Miyamoto, T. (1985). Mother to child transmission of human T-cell leukemia virus type I. *Gan No Rinsho*, **76**, 474–80.

Hrdy, D. B. Carlson, J. R., Yee, J. L. & Armstrong, M. Y. K. (1989). Need to confirm HTLV-I screening assays. *Lancet*, **i**, 109.

Jenkins, C. L., (1988). Health in the early contact period: a contemporary example from Papua New Guinea. *Social Science & Medicine*, **26**, 997–1006.

Jenkins, C., Dimitrakakis, M., Cook, I., Sanders, R. & Stallman, N. (1989). Culture change and epidemiological patterns among the Hagahai, Papua New Guinea. *Human Ecology*, **17**, 27–57.

Kazura, J. W., Saxinger, W. C., Wenger, J., Forsyth, K., Lederman, M. M., Gillespie, J. A., Carpenter, C. C. J. & Alpers, M. P. (1987). Epidemiology of human T-cell leukemia virus type I infection in East Sepik Province, Papua New Guinea. *Journal of Infectious Diseases*, **155**, 1100–1107.

Kinoshita, K., Hino, S., Amagasaki, T., Ikeda, S., Yamada, Y., Suzuyama, J., Momita, S., Toriya, K., Yamihira, S. & Ichimaru, M. (1984). Demonstration of adult T-cell leukemia virus antigen in milk from three seropositive mothers. *Gan No Rinsho*, **75**, 103–105.

Lillehoj, E. P., Alexander, S. S., Tai, C.-C., Dubrule, C. J., Adams, R., Manns, A., Wiktor, S. Z., Blattner, W. A., Cyrus, S., Decker, A. & Swenson, S. (1990).

Development and evaluation of an HTLV-I serologic confirmatory assay incorporating a recombinant envelope polypeptide. *Journal of Clinical Microbiology*, **28**, 2653–58.

Osame, M., Matsumoto, M., Usuku, K., Izumo, S., Ijichi, N., Amitani, H., Tara, M. & Igata, A. (1987). Chronic progressive myelopathy associated with elevated antibodies to human T-lymphotropic virus type I and adult T-cell leukemia-like cells. *Annals of Neurology*, **21**, 117–122.

Rodgers-Johnson, P., Gajdusek, D. C., Morgan, O. S. t. C., Zaninovic, V., Sarin, P. & Graham, D. S. (1985). HTLV-I and HTLV-III antibodies in tropical spastic paraparesis. *Lancet*, **ii**, 1247–1248.

Rodgers-Johnson, P., Garruto, R. M. & Gajdusek, D. C. (1988). Tropical myeloneuropathies – a new aetiology. *Trends in Neuroscience*, **11**, 526–532.

Rodgers-Johnson, P., Garruto, R. M., Yanagihara, R. & Gajdusek, D. C. (1990). Human T-lymphotropic virus type I: a retrovirus causing chronic myeloneuropathies in tropical and temperate climates. *American Journal of Human Biology*, **2**, 429–438.

Roman, G. C. (1987). Tropical spastic paraparesis in the Seychelle Islands: a clinical and case-control neuroepidemiologic study. *Neurology*, **37**, 1323–1328.

Tournier-Lasserve, E., Gout, O., Gessain, A., Iba-Zizen, M. T., Lyon-Caen, O., Lhermitte F., de The, G. (1987). HTLV-I, brain abnormalities on magnetic resonance imaging, and relation with multiple sclerosis. *Lancet*, **ii**, 49–50.

Weber, J. N., Banatvala, N., Clayden, S., McAcam, K. P.W. I., Palmer, S., Moulsdale, H., Tosswill., J., Dilger, P., Thorpe, R. & Amann, S. (1989). HTLV-I infection in Papua New Guinea: evidence for serologic false positivity. *Journal of Infectious Diseases*, **159**, 1025–1028.

Yanagihara, R., Ajdukiewicz, A. B., Garruto, R. M., Sharlow, E. R., Wu, X. Y., Alemaena, O., Sale, H., Alexander, S. S. & Gajdusek D. C. (1991b). Human T lymphotropic virus type I infection in the Solomon Islands. *American Journal of Tropical Medicine and Hygiene*, **44**, 122–130.

Yanagihara, R., Ajdukiewicz, A. B., Nerurkar, V. R., Garruto, R. M. & Gajdusek, D. C. (1991c). Verification of HTLV-I infection in the Solomon Islands by virus isolation and gene amplification. *Japanese Journal of Cancer Research*, **82**, 240–244.

Yanagihara, R., Garruto, R. M., Miller, M. A., Leon-Monzon, M., Liberski, P. P., Gajdusek, D. C., Jenkins, C. L., Sanders, R. C. & Alpers, M. P. (1990b). Isolation of HTLV-I from members of a remote tribe in New Guinea. *New England Journal of Medicine*, **323**, 993–994.

Yanagihara, R., Jenkins, C. L., Alexander, S. S., Mora, C. A. & Garruto, R. M. (1990a). HTLV-I infection in Papua New Guinea: high prevalence among the Hagahai confirmed by Western analysis. *Journal of Infectious Diseases*. **162**, 649–654.

Yanagihara, R., Nerurkar, V. R., Ajdukiewicz, A. B. (1991d). Comparison between strains of human T lymphotropic virus type I isolated from inhabitants of the Solomon Islands and Papua New Guinea. *Journal of Infectious Diseases*, **164**, 443–449.

Yanagihara, R., Nerurkar, V. R., Garruto, R. M., Miller, M. A., Leon-Monzon, M. E., Jenkins, C. L., Sanders, R. C., Liberski, P. P., Alpers, M. P. & Gajdusek, D. C. (1991a). Characterization of a variant of human T lymphotropic virus type I isolated from a member of a remote, recently contacted group in Papua New Guinea. *Proceedings of the National Academy of Sciences USA*, **88**, 1446–1450.

Zaninovic, V. (1987). Tropical spastic paraparesis. *Lancet*, **ii**, 280.

15 *Analysis of genes associated with hypercholesterolaemia in the Japanese population*

HIDEO HAMAGUCHI, YUKA WATANABE, YASUKO YAMANOUCHI, HISAKO YANAGI, TADAO ARINAMI, RYUNOSUKE MIYAZAKI, SHIGERU TSUCHIYA AND KIMIKO KOBAYASHI

Genes associated with hypercholesterolaemia have been detected and characterised at the loci for low density lipoprotein (LDL) receptor, apolipoprotein B, apolipoprotein E, and apolipoprotein (a) (Goldstein & Brown, 1989; Soria *et al.*, 1989; Mahley & Rall, 1989; Utermann, 1989). Among them, genes at the loci for apolipoprotein E and apolipoprotein (a) are polymorphic and act as polymeric genes for multifactorial hypercholesterolaemia. These polymorphic genes have been identified by the analysis of proteins using electrophoretic methods. It is interesting to examine whether these polymorphic and seemingly deleterious genes are identical at the DNA and amino acid sequence levels among different ethnic groups. On the other hand, deleterious genes detected at the loci for LDL receptor and apolipoprotein B are not polymorphic but act as major genes which cause autosomal dominant hypercholesterolaemia. These genes have been shown to be a cause of premature coronary heart disease (Goldstein & Brown, 1989; Soria *et al.*, 1989). The data obtained by the analysis of DNA suggest the same mutational origin for some of these deleterious genes (Goldstein & Brown, 1989) and their subsequent spread in populations.

This communication reports the results of molecular genetic studies on the phenotype E4 determined by the alleles at the locus for apolipoprotein E and on familial hypercholesterolaemia caused by the mutant LDL receptor gene in the Japanese population.

Apolipoprotein E4 in the Japanese population

Apolipoprotein E is a protein constituent of plasma lipoproteins and plays an important role in lipoprotein metabolism (Mahley,

Table 15.1. ***Frequencies of apolipoprotein E polymorphism alleles in different populations***

Country	*N*	*ε2*	*ε3*	*ε4*	*Reference*
Finland	1577	0.039	0.767	0.194	Lehtimäki *et al.*, 1990
Finland	615	0.041	0.733	0.227	Ehnholm *et al.*, 1986
France	434	0.123	0.751	0.126	Davignon *et al.*, 1988
Germany	1031	0.077	0.773	0.150	Utermann *et al.*, 1982
Germany	1000	0.078	0.783	0.139	Menzel *et al.*, 1983
Scotland	400	0.080	0.770	0.150	Cumming & Robertson, 1984
USA	1209	0.075	0.786	0.135	Ordovas *et al.*, 1987
New Zealand	426	0.120	0.720	0.160	Wardell *et al.*, 1982
Netherlands	2018	0.082	0.750	0.167	Smit *et al.*, 1988
Japan	608	0.049	0.841	0.095	present study

1988). The genetic polymorphism of apolipoprotein E is under the control of three common alleles (*ε2*, *ε3*, *ε4*) that specify isoforms apolipoprotein E2, E3 and E4 respectively (Mahley & Rall, 1989). The molecular basis for the apolipoprotein E polymorphism was established by Mahley and coworkers (Weisgraber *et al.*, 1981; Rall *et al.*, 1982; Paik *et al.*, 1985). The apolipoproteins E2, E3, and E4 differ from one another by single amino acid substitutions (Mahley & Rall, 1989). Table 15.1 shows the reported gene frequencies in apolipoprotein E polymorphisms in populations in which 400 or more subjects were analysed. Since the *ε3* allele is by far the most common apolipoprotein, E3 is considered to be the parent form of apolipoprotein E. While homozygosity for apolipoprotein E2 is the primary molecular defect in type III hyperlipoproteinaemia, apolipoprotein E2 is usually associated with lower plasma cholesterol levels (Mahley & Rall, 1989). On the other hand, it has been reported that apolipoprotein E4 is associated with higher plasma cholesterol levels (Mahley & Rall, 1989). In addition, an association of apolipoprotein E4 with hypercholesterolaemia has been shown in children as well as in adults (Utermann *et al.*, 1982; Leren *et al.*, 1985; Yanagi *et al.*, 1990).

Using two-dimensional gel electrophoresis, apolipoprotein E phenotypes were examined in 604 apparently healthy Japanese male adults who visited a hospital in Tokyo for medical checkup. The subjects were not excluded by any criterion other than the

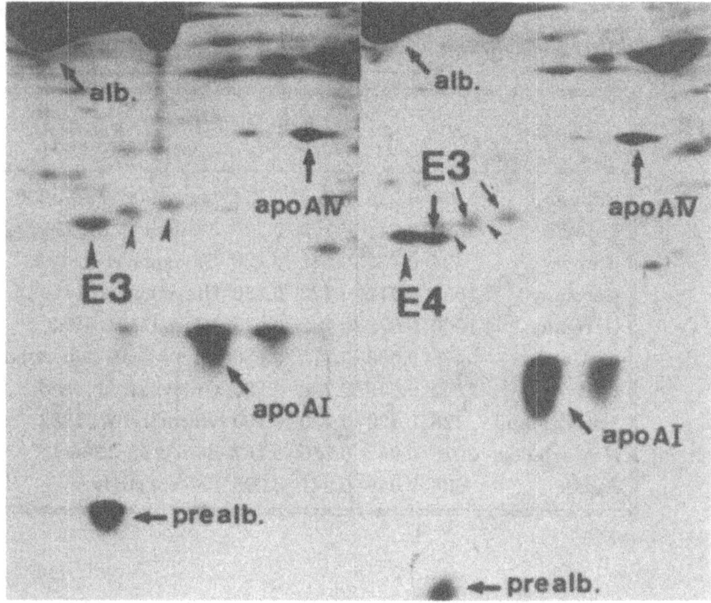

Figure 15.1. Apolipoprotein E phenotypes 3/3 (left) and 3/4 (right) in two-dimensional electrophoresis. E3 and E4 indicate apolipoprotein E3 and E4 isoforms, respectively. Isoelectric focusing was from left to right, and molecular weight separation by SDS-polyacrylamide gel electrophoresis was from top to bottom.

presence of diabetes mellitus, abnormal thyroid function or abnormal kidney function. Figure 15.1 shows the phenotypes of apolipoprotein E3/3 and E3/4 detectable with two-dimensional gel electrophoresis (Yanagi *et al.*, 1990). The apolipoprotein E phenotypes observed are shown in Table 15.2 and the frequencies of the *ε2*, *ε3* and *ε4* alleles are included in Table 15.1. In Japanese the frequency of the *ε3* allele is the highest. The frequency of the *ε2* allele is similar to that in the population of Finland, but is much lower than in other populations. The frequency of the ε4 allele in Japanese is much lower than in the population in Finland and also seems to be lower than in other populations. About 18% of Japanese, however, possess the *ε4* allele.

Table 15.3 shows mean serum cholesterol levels and the prevalence of hypercholesterolaemia (serum cholesterol ≥ 240mg/dl) in subjects with apolipoprotein E phenotypes 3/3 and 3/4. As expected, mean serum cholesterol levels were higher in

Table 15.2. *Apolipoprotein E phenotype distribution in a Japanese population*

Phenotype	*3/3*	*3/4*	*2/3*	*2/4*	*3/7*	*Other*	*Total*
N (%)	432 (71.1)	96 (15.8)	50 (8.2)	9 (1.5)	9 (1.5)	12 (2.0)	608

Subjects are Japanese males who ranged in age from 24 to 65 years with an average age of 47.6. Others in apolipoprotein E phenotype include four with type 4/4, four with type 3/5, two with type 4/7, one with type 4/5, and one with type 2/5.

Table 15.3. *Mean serum cholesterol levels and prevalence of hypercholesterolaemia in apolipoprotein E phenotypes 3/3 and 3/4.*

Phenotype	*N*	*Cholesterol levels*		*Hyper-cholesterolaemia (≥240mg/dl)*	
		Age (mean)	*(mean±SD) (mg/dl)*	*N*	*(%)*
3/3	432	48.0	189.6+31.4	29	(6.7)
3/4	96	46.5	200.7+32.0**	16*	(14.6)**

$^{**}t=3.08$ $p<0.005$, $^{*}\chi^2=6.54$ $p<0.02$

the subjects with apolipoprotein E 3/4 than in the subjects with apolipoprotein E3/3. The prevalence of hypercholesterolaemia was also higher in the subjects with apolipoprotein E3/4 than in those with apolipoprotein E3/3.

It is important to clarify whether the amino acid substitution in apolipoprotein E4 in Japanese is the same as that in Caucasians. This subject was examined using polymerase chain reaction and dot blot hybridisation with allele-specific oligonucleotide probes. Figure 15.2 shows the sequences of apolipo-

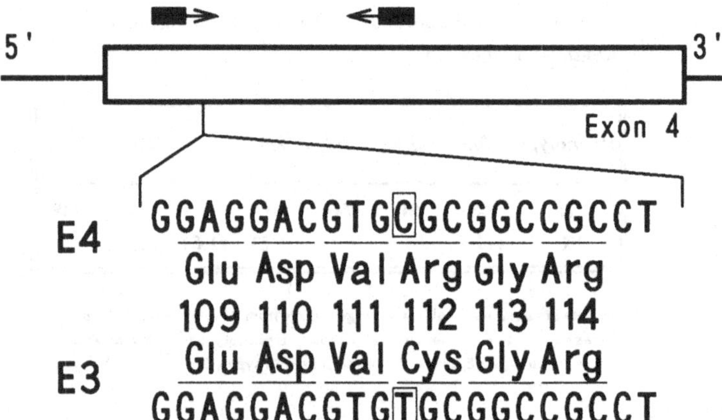

Figure 15.2. The DNA and amino acid sequences near the region of codon 112 in exon 4 of apolipoprotein E3 and E4. The T to C substitution in codon 112 which causes the Cys to Arg substitution in apolipoprotein E4 is shown by boxes. The DNA sequences represent the sense strands. The sequence (19-mer) of the anti-sense strand complementary to each sense strand was used as the probes for the ε3 allele and ε4 allele, respectively. The positions of primers are shown by the solid bars. The amplified region of the apolipoprotein E gene is bases 3690-4020 in exon 4.

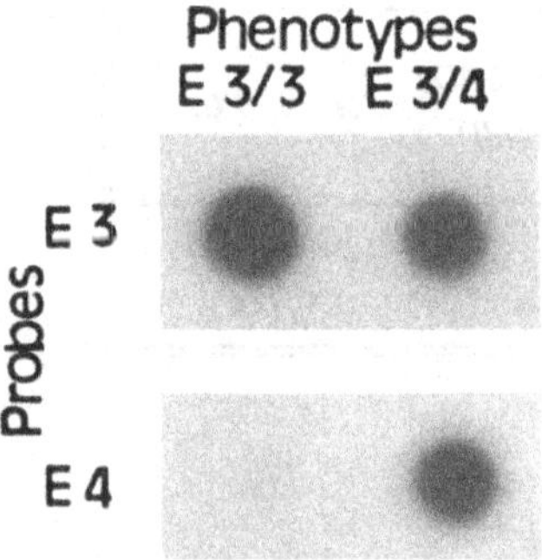

Figure 15.3. Dot blot hybridisation of amplified DNA of the apolipoprotein E gene derived from subjects with apolipoprotein E phenotype 3/3 and 3/4. The amplified DNA from the subject with the phenotype 3/4 hybridises with both the probes for the ε3 and ε4 alleles.

protein E3 and E4 near the region of exon 112 in the apolipoprotein E gene reported by Mahley and coworkers (Weisgraber *et al.*, 1981; Rall *et al.*, 1982; Paik *et al.*, 1985). Figure 15.2 also presents the sequences of the oligonucleotide

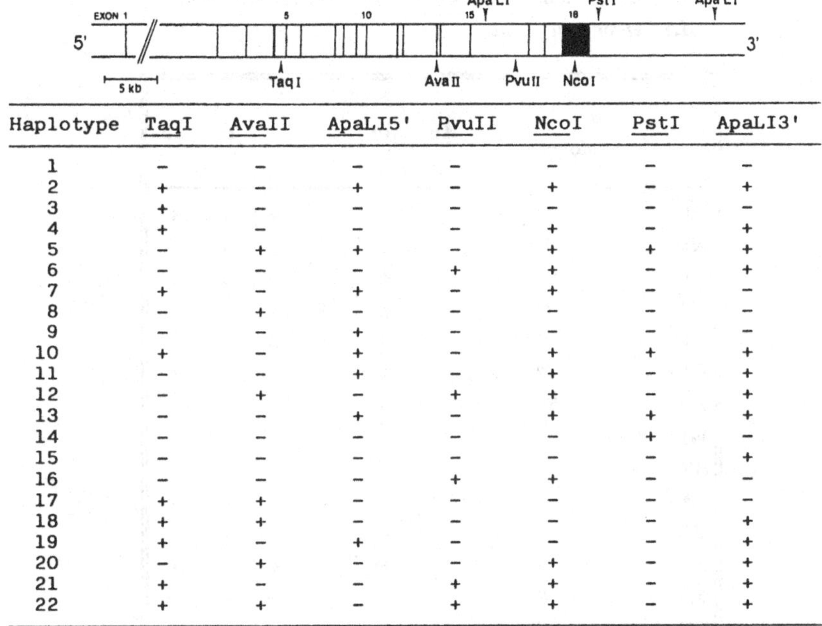

Haplotype	TaqI	AvaII	ApaLI5'	PvuII	NcoI	PstI	ApaLI3'
1	-	-	-	-	-	-	-
2	+	-	+	-	+	-	+
3	+	-	-	-	-	-	-
4	+	-	-	-	+	-	+
5	-	+	+	-	+	+	+
6	-	-	-	+	+	-	+
7	+	-	+	-	+	-	-
8	-	+	-	-	-	-	-
9	-	-	+	-	-	-	-
10	+	-	+	-	+	+	+
11	-	-	+	-	+	-	+
12	-	+	-	+	+	-	+
13	-	-	+	-	+	+	+
14	-	-	-	-	-	+	-
15	-	-	-	-	-	-	+
16	-	-	-	+	+	-	-
17	+	+	-	-	-	-	-
18	+	+	-	-	-	-	+
19	+	-	+	-	-	-	+
20	-	+	-	-	+	-	+
21	+	-	-	+	+	-	+
22	+	+	-	+	+	-	+

Figure 15.4. Seven RFLPs and 22 RFLP haplotypes on the LDL receptor locus observed in Japanese. The map positions of the polymorphic restriction sites are indicated by the arrows. The presence or absence of each of the polymorphic restriction sites for each of the haplotypes is shown as + and - symbols.

probes for apolipoprotein E3 and E4 used in the present study and the amplified region of exon 4 in the apolipoprotein E gene. Twenty five *ε4* alleles derived from unrelated Japanese were examined; all the *ε4* alleles hybridised with the oligonucleotide probe specific for apolipoprotein E4. Figure 15.3 shows a representative result of dot blot hybridisation with the allele-specific oligonucleotide probes. These data indicate that the sequence near the region of codon 112 is identical with that in Caucasians in most, if not all, of the *ε4* alleles in Japanese and that the amino acid substitution Cys → Arg is also present at residue 112 in apolipoprotein E4 in Japanese. It is important to enquire whether the mutational origin of the *ε4* allele is the same in ethnically different populations.

Familial hypercholesterolaemia

The level of cholesterol in human plasma is to a large extent

Table 15.4. ***Deletions and RFLP haplotypes of the mutant LDL receptor genes in Japanese***

Family	*RFLP haplotypes*	*T*	*Av*	*A5*	*Pv*	*N*	*P*	*A3*
OF	D	+	-	d	d	+	-	+
NM	D	-	-	d	d	d	d	d
YI	D^5	-	+	+	-	+	+	+
ST	D	-	-	d	d	d	-	-
KM	D	-	-	d	d	d	-	-
SS	T^2	+	-	+	-	+	-	+
KS	T^2	+	-	+	-	+	-	+
WI	1	-	-	-	-	-	-	-
SY	1	-	-	-	-	-	-	-
TA	1	-	-	-	-	-	-	-
YK	1	-	-	-	-	-	-	-
KT	2	+	-	+	-	+	-	+
FH	2	+	-	+	-	+	-	+
TN	5	-	+	+	-	+	+	+
OT	5	-	+	+	-	+	+	+
OS	5	-	+	+	-	+	+	+
OK	6	-	-	-	+	+	-	+
SH	7	+	-	+	-	+	-	-
SM	12	-	+	-	+	+	-	+
TT	13	-	-	+	-	+	+	+
MH	16	-	-	-	+	+	-	-

The presence or absence of the restriction site is represented by + or -, respectively. D indicates the mutant gene with a partial deletion and D^5 means a partial deletion associated with the haplotype 5. Deletions at restriction enzyme sites are shown by d. T^2 means the presence of the abnormal *Taq*I band associated with the haplotype 2.

controlled by the interaction of LDL with LDL receptor. Defects of the LDL receptor gene cause familial hypercholesterolaemia (Goldstein & Brown, 1989). The heterozygote frequency of familial hypercholesterolaemia is estimated to be 1 in 500 persons in Japanese as in most Caucasian populations (Mabuchi *et al.*, 1979). Since the fitness of the heterozygote for familial hypercholesterolaemia is more than 0.95, the mutant LDL receptor genes seem to be transmitted with little loss from generation to

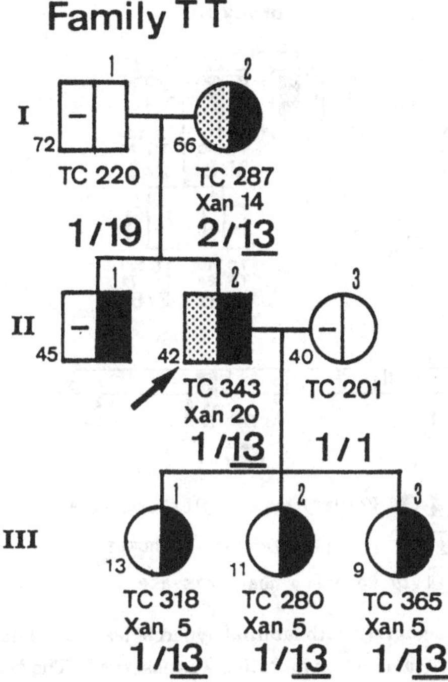

Primary hypercholesterolemia

Achilles tendon xanthomas

Coronary heart disease

Figure 15.5. A pedigree with familial hypercholesterolaemia. The underlined numerals indicate the RFLP haplotype of the LDL receptor gene co-segregating with hypercholesterolaemia. Tc = total cholesterol levels.

generation in most pedigrees. In general, the origin of the mutant LDL receptor genes seems to differ in different pedigrees. In some ethnic groups, however, it has been suggested that the mutant genes with the same mutational origin spread through the populations (Goldstein & Brown, 1989).

In our molecular genetic studies of the mutant LDL receptor genes of familial hypercholesterolaemia (Yamakawa *et al.*, 1988, 1989, 1991), restriction fragment length polymorphisms at the LDL receptor locus were examined in Japanese. Figure 15.4 shows the sites of seven RFLPs detected that have high

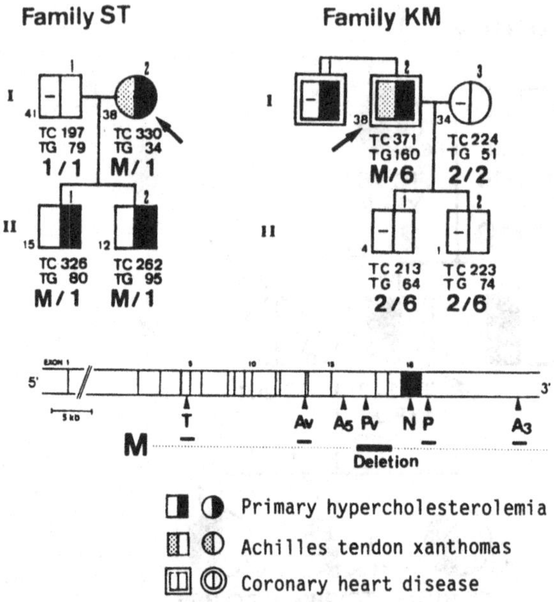

Figure 15.6. Two pedigrees with familial hypercholesterolaemia in which an apparently identical deletion mutation was observed. The M symbol indicates the presence of the deletion mutation in the LDL receptor gene. The deleted region in the LDL receptor gene is shown.

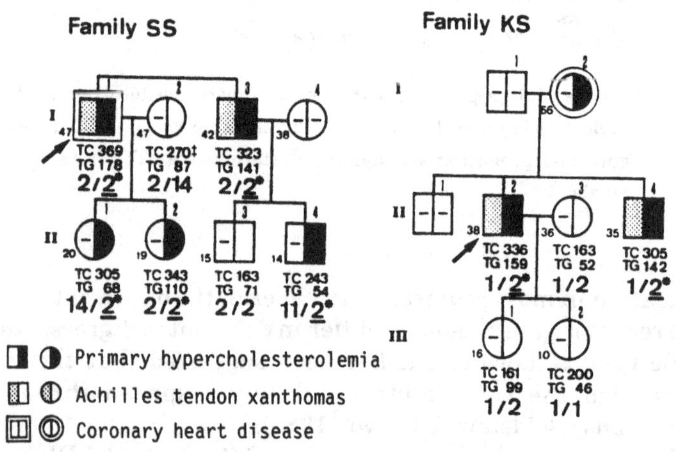

Figure 15.7. Two pedigrees with familial hypercholesterolaemia in which an identical abnormal *Taq*I band associated with the RFLP haplotype 2 co-segregated with hypercholesterolaemia. The asterisk indicates the presence of the abnormal *Taq*I band.

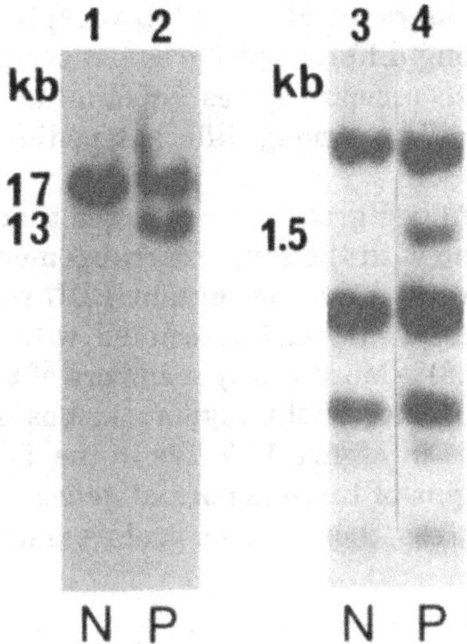

Figure 15.8. An abnormal band derived from the mutant LDL receptor gene with a partial deletion of about 4kb observed in hypercholesterolaemic members of families in Figure 15.6 (left, 13kb band in lane 2), and an abnormal *Taq*I band detected in the LDL receptor gene in hypercholesterolaemic members of families in Figure 15.7 (right, 1.5 kb band in lane 4). N = normolipidemic subject; P = hypercholesterolaemic subject.

polymorphism information content in Japanese. Thus far 22 RFLP haplotypes have been detected among 121 normal LDL receptor genes by family studies of 37 different pedigrees (Figure 15.4).

Twenty one pedigrees with familial hypercholesterolaemia were analysed by Southern blotting using an LDL receptor cDNA. Hypercholesterolaemia co-segregated with gross rearrangement, RFLP haplotype, or an abnormal *Taq*I band in the 21 pedigrees. Co-segregated genetic markers of the LDL receptor gene with hypercholesterolaemia included four different deletion mutations, eight different RFLP haplotypes and one *Taq*I abnormal band (Table 15.4). Figures 15.5-7 show the representative families. Figure 15.8 presents Southern blots of the gross rearrangement in the LDL receptor gene observed in the families in Figure 15.6 and an abnormal *Taq*I band observed in the families in Figure 15.7.

Since co-segregated genetic markers of the LDL receptor gene have a tendency to vary among different pedigrees, it seems that the origin of the mutant LDL receptor genes for familial hypercholesterolaemia generally differs among different pedigrees in Japanese.

In this study, two different pedigrees were observed that share the mutant LDL receptor gene with the same rearrangement, and another two different pedigrees share the mutant LDL receptor gene with the same abnormal *Taq*I band associated with RFLP haplotype 2 (Figures 15.6-15.8). Most family members of the two pedigrees in Figure 15.6 live in the Tohoku region and most family members of the two pedigrees in Figure 15.7 live in the Tyugoku region. The mutational origins of these abnormal genes seem to be different in different pedigrees, though some pedigrees have the same mutant in common.

Conclusion

The phenotype E4 of apolipoprotein E is present in about 18% of Japanese and associated with higher serum cholesterol levels and hypercholesterolaemia. The analysis of the ε4 allele by dot blot hybridisation with allele-specific oligonucleotide probes revealed that the amino acid substitution Cys → Arg at residue 112 is present in most, if not all, of the apolipoprotein E4 molecules in Japanese as in Caucasians. It is important to clarify whether the mutational origin of the ε4 allele is the same in Japanese and Caucasians. On the other hand, the analysis of the mutant LDL receptor genes responsible for familial hypercholesterolaemia by Southern blotting revealed that the mutational origins of the mutant genes tend to be different among different pedigrees in Japanese. It is suggested, however, that some of the mutant LDL receptor genes spread in apparently unrelated pedigrees with familial hypercholesterolaemia.

References

Cumming, A. M. & Robertson, F. W. (1984). Polymorphism at the apoprotein-E locus in relation to risk of coronary disease. *Clinical Genetics*, **25**, 310–313.

Davignon, J., Gregg, R. E. & Sing, C. F. (1988). Apolipoprotein E polymorphism and atherosclerosis. *Arteriosclerosis*, **8**, 1–21.

Ehnholm, C., Lukka, M., Kuusi, T., Nikkilä, E. & Utermann, G. (1986). Apolipoprotein

E polymorphism in the Finnish population: gene frequencies and relation to lipoprotein concentrations. *Journal of Lipid Research*, **27**, 227–235.

Goldstein, J. L. & Brown, M. S. (1989). Familial hypercholesterolemia. In *The metabolic basis of inherited disease*, 6th edn, ed. C. R. Scriver, A. L. Beaudet, W. S. Sly, D. Valle, pp. 1215–1250. New York: McGraw-Hill.

Lehtimäki, T., Moilanen, T., Viikari, J., Åkerblom, H. K., Ehnholm, C., Rönnemaa, T., Marniemi, J., Dahlen, G. & Nikkari, T. (1990). Apolipoprotein E phenotypes in Finnish youths: a cross-sectional and 6-year follow-up study. *Journal of Lipid Research*, **31**, 487–495.

Leren, T. P., Borresen, A. L., Berg, K., Hjermann, I. & Leren, P. (1985). Increased frequency of the apolipoprotein E-4 isoform in male subjects with multifactorial hypercholesterolemia. *Clinical Genetics*, **27**, 458–462.

Mabuchi, H., Tatami, R., Ueda, K., Ueda, R., Haba, T., Kametani, T., Watanabe, A., Wakasugi, T., Itoh, S., Koizumi, J., Ohta, M., Miyamoto, S. & Takeda, R. (1979). Serum lipid and lipoprotein levels in Japanese patients with familial hypercholesterolemia. *Atherosclerosis*, **32**, 435–444.

Mahley, R. W. (1988). Apolipoprotein E: cholesterol transport protein with expanding role in cell biology. *Science*, **240**, 622–630.

Mahley, R. W. & Rall, S. C. (1989). Type III hyperlipoproteinemia (Dysbetalipoproteinemia). In: *The metabolic basis of inherited disease*, 6th edn. pp. 1195–1213. ed. C. R. Scriver, A. L. Beaudet, W. S. Sly, & D. Valle, New York: McGraw-Hill.

Menzel, H. J., Kladetzky, R. G. & Assmann, G. (1983). Apolipoprotein E polymorphism and coronary artery disease. *Arteriosclerosis*, **3**, 310–315.

Ordovas, J. M., Litwack-Klein, L., Wilson, P. W. F., Schaefer, M. M. & Schaefer, E. J. (1987) Apolipoprotein E isoform phenotyping methodology and population frequency with identification of apoE1 and apoE5 isoforms. *Journal of Lipid Research*, **28**, 371–380.

Paik, Y-K, Chang, D. J., Reardon, C. A., Davies, G. E., Mahley, R. W. & Taylor, J. M. (1985). Nucleotide sequence and structure of the human apolipoprotein E gene. *Proceedings of the National Academy of Sciences USA*, **82**, 3445–3449.

Rall, S. C., Weisgraber, K. H. & Mahley, R. W. (1982). Human apolipoprotein E. The complete amino acid sequence. *Journal of Biological Chemistry*, **257**, 4171–4178.

Smit, M., de Knijff, P., Rosseneu, M., Bury, J., Klasen, E., Frants, R. & Havekes, L. (1988). Apolipoprotein E polymorphism in the Netherlands and its effect on plasma lipid and apolipoprotein levels. *Human Genetics*, **80**, 287–292.

Soria, L. F., Ludwig, E. H., Clarke, H. R. G., Vega, G. L., Grundy, S. M. & McCarthy, B. J. (1989). Association between a specific apolipoprotein B mutation and familial defective apolipoprotein B-100. *Proceedings of the National Academy of Sciences USA*, **86**, 587–591.

Utermann, G. (1989). The mysteries of lipoprotein(a). *Science*, **246**, 904–910.

Utermann, G., Steinmets, A. & Weber, W. (1982). Genetic control of human apolipoprotein E polymorphism: comparison of one- and two-dimensional techniques of isoprotein analysis. *Human Genetics*, **60**, 344–351.

Wardell, M. R., Suckling, P. A. & Janus, E. D. (1982). Genetic variation in human apolipoprotein E. *Journal of Lipid Research*, **23**, 1174–1182.

Weisgraber, K. H., Rall, S. C. & Mahley, R. W. (1981). Human E apoprotein heterogeneity. Cysteine-arginine interchanges in the amino acid sequence of the apo-E isoforms. *Journal of Biological Chemistry*, **256**, 9077–9083.

Yamakawa, K., Okafuji, T., Iwamura, Y., Yuzawa, K., Satoh, J., Hattori, N., Yamanouchi, Y., Yanagi, H., Kawai, K., Tsuchiya, S., Russell, D. W. & Hamaguchi, H., (1988). TaqI polymorphism in the LDL receptor gene and a TaqI 1.5-kb band associated with familial hypercholesterolemia. *Human Genetics*, **80**, 1–5.

Yamakawa, K., Takada, K., Yanagi, H., Tsuchiya, S., Kawai, K., Nakagawa, S. & Hamaguchi, H. (1989). Three novel partial deletions of the low-density lipoprotein gene in familial hypercholesterolemia. *Human Genetics* **80**, 317–321.

Yamakawa, K. Yanagi, H., Saku, K., Sasaki, J., Okafuji, T., Shimakura, Y., Kawai, K.,

Tsuchiya, S., Takada, K., Naito, S., Arakawa, K. & Hamaguchi, H. (1991). Family studies of the LDL receptor gene of relatively severe hereditary hypercholesterolemia associated with Achilles tendon xanthomas. *Human Genetics*, **86**, 445–449.

Yanagi, H., Shimakura, Y., Yamanouchi, Y., Watanabe, Y., Tsuchiya, S. & Hamaguchi, H. (1990). Association of hypercholesterolemia and apolipoprotein E4 in school children. *Clinical Genetics*, **38**, 264–269.

16 *Migrant studies and their problems*

PAUL T. BAKER

Migrants are in themselves worthy of study by human biologists because they encounter a number of health and behavioural problems as a consequence of moving into what are frequently very alien natural and social environments. The study of the migration process also enhances our understanding of the consequences of the spread of our species over the globe. For most human population biologists however, be they geneticists, anthropologists, epidemiologists or physiologists, the basic reason for an avid interest in migrants is the opportunity that they provide for understanding the underlying causes for variability in our species.

Although the rapidly expanding field of molecular genetics is developing exciting new explanations of the ways in which particular genes in suitable environments produce specific phenotypes, and previously unimaginable documentation of the past and present flow of genes through population movement, it seems unlikely that in the foreseeable future such studies will provide an explanation for most group and individual variability in human biology and behaviour. In the meantime migrant groups provide natural experiments which can be used to explain how genes and environment interact to produce both individual and group differences. Indeed it is only in the context of studying phenotypic development in contrasting and altered environmental conditions that we will be able to predict genetic effects on the phenotype. Within the discipline of human biology the study of migrants was initially used primarily to determine whether such traits as head shape and body size were genetically fixed traits in what were then termed races. These studies led the way to the current understanding that almost all aspects of the phenotype are modified to varying degrees by the environments in which the genetic code is expressed (Lasker & Mascie-Taylor, 1988).

Despite these advantages of using migrant studies for understanding environmental effects on development, it must always be kept in mind that such studies can easily lead to mistaken conclusions when the specific causes for biological changes are sought. This occurs because it is impossible to control by research design for all of the potentially significant environmental variables which change as a result of migration. Therefore the focus of my cautionary comments will relate to the use of migrant studies for determining the causes for a specific phenotypic expression of an individual or group genotype.

While the early studies were not directly designed to explain the causes for the altered phenotype, many authors offered explanations based upon their particular scientific background. For example the larger average body size of first generation migrants who grew up in a new environment was often attributed to a better 'environment' (Bogin, 1988) while the altered morphology of second generation migrants compared to the first generation was attributed to such genetic causes as heterosis (Livi, 1896; Ammon, 1899). While these explanations for the differences may indeed prove to be valid for some of the findings, the designs of the studies and consequently the results certainly did not justify identifying such causes for changes.

The problems of applying simplistic deductive logic to the results for the study of migrants became increasingly apparent when migrant studies were used to explore the causes for such adult disorders as cardiovascular disease (CVD). Since the findings that some traditional populations had low blood pressures and low salt intakes while migrants from these populations to modern societies often developed hypertension, many researchers concluded that high salt intake caused the high prevalence of hypertension in populations living in modern societies. Conclusions about causation have also been drawn from migrant studies showing low fat intakes and cholesterol levels in traditional populations as compared with high fat intakes and high cholesterol levels in migrants. However such single-cause explanations for these precursors of CVD have not been verified by more detailed studies.

Additional examples of the problems inherent in the use of migrant studies for understanding the causes of biological variability in our species could be greatly amplified but I will illustrate more fully the need for caution with an example from our long-term study of Samoans. Our initial studies of Samoans in Hawaii and American Samoa showed that these populations probably had the highest body mass indices of any population in the world (Bindon, 1981). As the study proceeded we found that the Samoans living in traditional villages in Western Samoa did not manifest these unusually large weights although they were as tall as the Samoans in American Samoa and Hawaii (Pawson, 1986). Indeed the most traditional Samoan villagers had at almost all ages weights and heights which resembled the US and English norms for the 1960s (Baker, 1981). On this basis one might conclude that their nutrition was adequate. Even so, conventional nutritional theory suggested that the difference between the normal and overweight populations should relate either to differential food intake due to diet change or contrasting activity patterns which affected energy expenditure.

Our subsequent nutritional and activity studies on the groups failed to support these logical conclusions. Physical activity was indeed lower for American Samoans and Hawaii Samoans compared to traditional villagers (Schendel, 1989) but total caloric expenditure was similar because of the larger body weights of Samoans in American Samoa and Hawaii (Pearson, 1990). Caloric intakes appeared also to be comparable in the various groups or perhaps lower in the thinner groups. Even the possibility that Samoans from Hawaii and American Samoa were fatter as a result of eating calorically dense foods was not supported since the percentages of calories provided by fat, protein, and carbohydrates were similar for the three groups. Thus despite both survey and detailed small group studies we could not demonstrate the reasons for the unusually large body masses of Samoans in modernised societies (Baker & Hanna, 1986).

Based on the high birth weights and rapid weight gain of infant Samoans in all environments, I have suggested that they and perhaps Polynesians in general have a genetic predisposition to large body weights. Even if this is correct the social or environmental factors which cause Samoans to develop extreme body weights in a modern society remain unexplained.

I hope that these remarks will not be viewed as a condemnation of migration studies. As stated earlier, I believe that such studies remain one of the best available methods for understanding how genes and environment interact to produce the phenotype. Nevertheless I believe equally firmly that causality for the biological changes encountered among migrants has often been too simply deduced and that more rigorous research designs are needed. It appears to me that too often in the past the cause for a change in the biology of migrants has been accepted as proven based on its frequent repetition in the literature rather than the validity of the proof.

References

Ammon, O. (1899). Zur Anthropologie der Badener. Jena cited in Boas, F. (1922). Report of an anthropometric investigation of the population of the United States. *Journal of the American Statistical Association*, **18**, 181–209.

Baker, P. T. (1981). Migration and human adaptation. In *Migration, adaptation and health in the Pacific*, ed. C. Fleming & I. Prior. Epidemiology Unit, Wellington, New Zealand.

Baker, P. T. & Hanna, J. M. (1986). Perspectives on health and behaviour of Samoans. In *The changing Samoans*, ed. P. Baker, J. Hanna & T. Baker. New York: Oxford University Press.

Bindon, J. R. (1981). Genetic and environmental influences on the morphology of Samoan adults. Ph. D. thesis, The Pennsylvania State University.

Bogin, B. (1988). Rural-to-urban migration. In *Biological aspects of human migration*, ed. C. G. N. Mascie-Taylor & G. W. Lasker. Cambridge: Cambridge University Press.

Lasker, G. W. & Mascie-Taylor, C. G. N. (1988). The framework of migration studies. In *Biological aspects of human migration*, ed. C. G. N. Mascie-Taylor & G. W., Lasker. Cambridge. Cambridge University Press.

Livi, R. (1896). *Antropometrica Militare*. Risultati ottenuti dello spoglio dei fogli sanitaru dei militari delle classi 1859–63. Rome.

Pawson, I. G. (1986). The morphological characteristics of Samoan adults. In *The changing Samoans*, ed. P. Baker, J. Hanna & T. Baker. New York: Oxford University Press.

Pearson, J. D. (1990). Estimation of energy expenditure in Western Samoa, American Samoa, and Honolulu by Recall Interviews and Direct Observation. *American Journal of Human Biology*, **2**, 313.

Schendel, D. E. (1989). Sex differences in factors associated with body fatness in Western Samoans. Ph. D. thesis. The Pennsylvania State University.

17 *Tokelau: migration and health in a small Polynesian society - a longitudinal study*

IAN PRIOR

Introduction

The Tokelau Island Migrant Study was established in 1968 as a multidisciplinary study to test hypotheses relating to the process of change associated with migration, including factors influencing changing blood pressure patterns with age. In its broadest sense it is a study of the relationship between social change and health in this small Polynesian society (Prior *et al.*, 1974). This has involved bringing together findings from three different scientific disciplines: ethnography, quantitative sociology, and epidemiology including genetic epidemiology. This report presents a brief review of the development of the project since it was commenced in 1968 together with a summary of some of the results of the cross-sectional and longitudinal analyses, particularly as regards body weight, diabetes, gout, hypertension and blood pressure. This synthesis outlines some of the dynamics and consequences of social change for individuals and populations. The study recognises that the migration process is complex and that it is taking place against the developing pattern of social change in two societies: Tokelau, the one from which the migrants originated, and New Zealand, the host country in which most of the migrants settled.

Tokelau is very small and, in geopolitical terms, isolated and insignificant. The three small atolls of Tokelau (Fakaofo, Nukunonu and Atafu) lie 480km north of Samoa and 3,200km from New Zealand (Figure 17.1). In 1966 there were 1901 Tokelauans in Tokelau and 445 in New Zealand. A hurricane in

Acknowledgements: The material in this chapter is being published more fully by Oxford University Press (1992) as 'Migration and Health in a Small Society: The Tokelau Experience', edited by A.F. Weesen, A. Hooper, J. Huntsman, I.A.M. Prior and C.E. Salmond. Acknowledgement is made to Orlando Press, the American Journal of Epidemiology and the Journal of Chronic Diseases, for permission to produce Figures 17.6, 17.7 and 17.8.

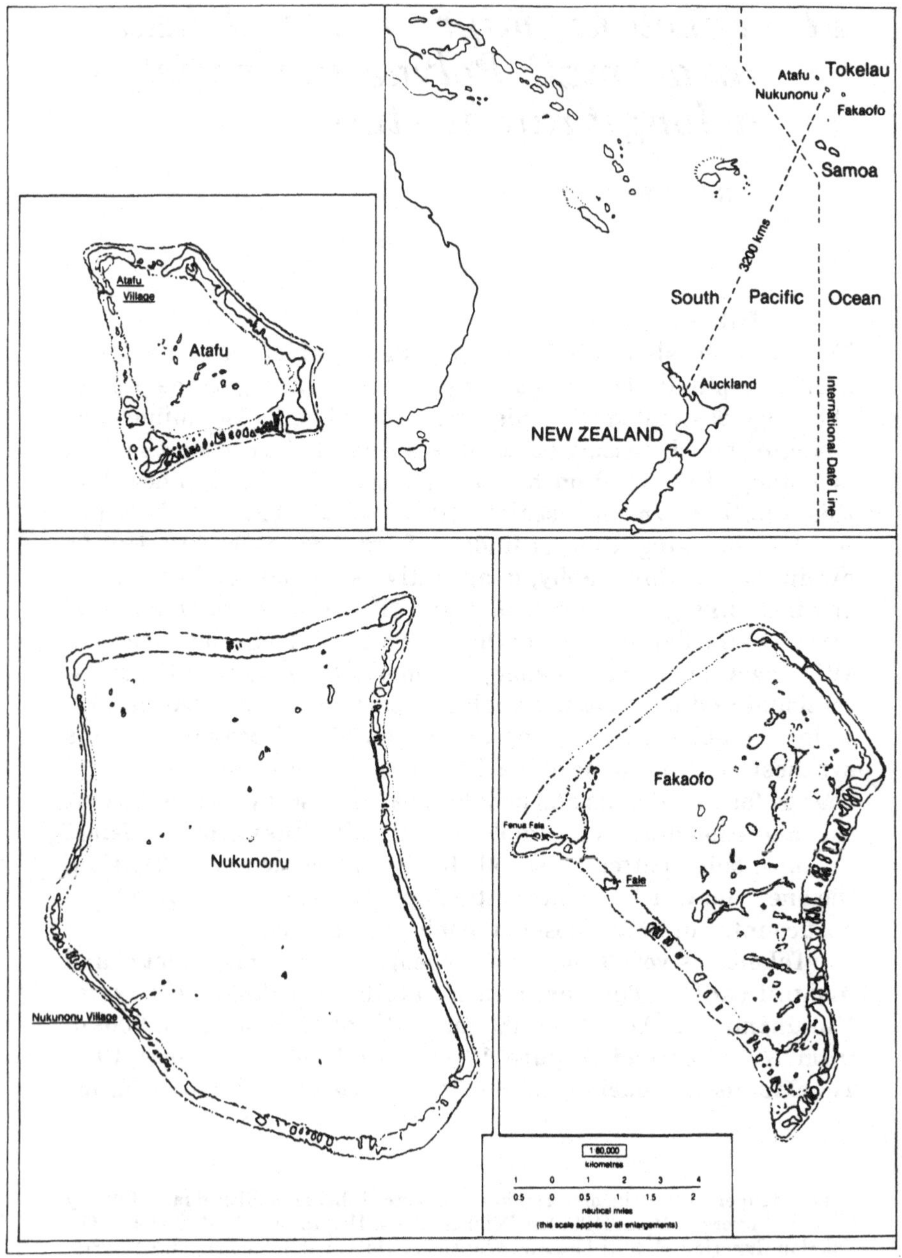

Figure 17.1. Map showing three Tokelau atolls, Fakaofo, Nukunonu and Atafu, and distance from New Zealand. Made available by Volunteer Service Abroad/Te Tuao Tawahi, New Zealand.

Tokelau led to the establishment of the Tokelau Resettlement Programme by the New Zealand Government bringing Tokelauans to New Zealand. This followed from the fact that the islanders had been given rights as New Zealand citizens in 1948 and that Tokelau was a New Zealand Dependency under the United Nations. In 1971, the total Tokelauan population (apart from a small number in Western and American Samoa, Australia and Hawaii) comprised only 3,471 persons, of whom half (52%) were already living in New Zealand. By 1982, it numbered 4,118, of whom about two thirds (62%) were in New Zealand. Over the whole period of the study, data were gathered from 5,011 persons, 843 of whom had died or been lost to follow-up by 1984. There has thus been a high measure of participation. An advantage of the study was the fact that a considerable number of the migrants had been examined in Tokelau prior to migration.

On Tokelau, the small Polynesian atoll, the subsistence society has a very limited economic resource base. It is dependent to an increasing extent on support from New Zealand and represents a major contrast with urban New Zealand. The documented considerable changes that have taken place in Tokelau society over the period of the study provide information on the process of change in such societies which may be relevant to other small Pacific populations.

The stimulus for the Tokelau study came from earlier studies in the Pacific of populations in whom blood pressure and weight showed little increase with age, such as Pukapuka in the Northern Cook Islands, a pattern in contrast to the considerable increase with age seen in Rarotonga in the Southern Cook Islands, New Zealand Maori and Europeans (Prior *et al.*, 1966). A study in Tokelau in 1963 showed that their blood pressures increased only slightly with age, rather similar to Pukapuka (Elliot, 1963). This strongly influenced the decision to set up the Tokelau Island Migrant Study as a longitudinal project in 1968. It was considered that the stresses of adapting to the process of migration and modernisation could have measurable physiological effects contributing to hypertension and other conditions as part of the cost of modern urban living. Other hypotheses related to the way in which changing life styles might influence the development of a range of specific disease conditions. The life-style changes included alterations in diet, increased salt intake, overnutrition as a result of increased prosperity and purchasing power, changing patterns of habitual exercise and fitness, use of alcohol and tobacco, changing social involvement and cultural support systems.

This is essentially the epidemiological 'risk factor' approach to the migrant experience. This contrasts with the more 'holistic' approach adopted by the social scientists which emphasises cultural change and puts forward hypotheses as to how this could influence the adaptive responses to individuals and their physiological functioning. The Tokelau Island Migrant Study provided an ideal opportunity to explore the explanatory powers of the two appraches (Cassel, 1974).

The question of the relationship between the individual and society, in both Tokelau and New Zealand, a central part of the study, involved examining the connection between (1) the distinction between public and private culture (Goodenough, 1963), and (2) the basic sociological notion that the individual consequences of social change are mediated by the particular position of each individual in his community's social structure.

The migrants in New Zealand were faced with a broad range of changes of living patterns, social contacts, work exposures, food and diet, that thrust them into the modernisation process. At the same time the firm bonds of community, the conservative force of strongly-held Tokelau values and precepts, and the security of customs and habit in combination would be expected to lessen the impetus for change. As the study proceeded it became clear that those remaining in Tokelau also were facing increasing changes. Isolation decreased with more frequent boat visits. Some people began to move to a cash economy related to the increasing development of the Public Service which threatened aspects of their traditional economic base. More education and social mobility also contributed to the social changes as the younger people started questioning some of the traditional ways.

The formal survey data on the population of Tokelauans, both atoll-dwellers and migrants to New Zealand, are based largely upon three comprehensive surveys conducted over fourteen years (1968-1982). For some individual Tokelauans, quantitative observations were made in precursor surveys dating as far back as to 1963 and in follow-up surveys as late as 1984. In terms of the formal analysis, however, quantitative data are presented here largely from the six major surveys of Tokelauans in three periods: Round I, 1971 in Tokelau, 1972-1974 in New Zealand; Round II, 1976 in Tokelau, 1975-1977 in New Zealand; and Round III, 1982 in Tokelau, 1980-1981 in New Zealand. These quantitative surveys linked in with active ethnographic research conducted since 1967, both in Tokelau and among the Tokelauan migrants in New Zealand. The population data for the three Rounds 1971 to 1982 is set out in Table 17.1.

Table 17.1. *Tokelau Island migrant study population. Data for 1971 to 1982*

	Total seen this Round in Tokelau or NZ	*Those not seen who had been seen before*	*Eligible population ever seen to given round*	*Percent of total population covered (a)*
Round I				
Children	1656	8 (b)	1664	99.5
Adults	1814	25	1839	98.6
Adults & children	3470	33 (c)	3503	99.1
Round II				
Children	1862	128	1990	93.6
Adults	2029	197	2226	91.2
Adults & children	3891	325	4216	92.3
Round III				
Children	1927	419	2346	82.1
Adults	2191	474	2665	82.2
Adults & children	4118	893	5011	82.2

Round I: 1971 in Tokelau and 1972-1974 in New Zealand.
Round II: 1976 in Tokelau and 1974-1976 in New Zealand.
Round III: 1982 in Tokelau and 1980-1981 in New Zealand.
Adults: 15 years and over; Children: less than 15 years.

(a) Includes both eligibles and deaths.
(b) Children were not included in the 1968 data set only incompletely in data gathered before Round I.
(c) Includes persons for whom information was gathered on the Unit 1968 expeditions to Tokelau or from examinations of New Zealand migrants between 1966 and 1970.

Changes in demographic patterns in Tokelau and New Zealand

Based on data from the Migrant Study files there were only 63 Tokelauans in New Zealand in 1962, increasing to 445 in 1966 when the New Zealand Government set up the Resettlement Programme to bring Tokelau young people and families to New Zealand. At the same time increasing numbers of young people were brought to New Zealand for schooling and trade training. By 1973 the tempo of migration had stepped up very considerably and there were at least 2,000 Tokelauans who had come to New Zealand under the Resettlement Programme or by a chain migration process with the assistance of their families.

During the later 1970s and the 1980s the flow of migrants subsided to low levels and most of the continuing increase in Tokelauans in New Zealand was contributed by the high birth rate of the migrant families. By 1985 a project census of Tokelauans living in New Zealand gave a total of 3,368. During the study medical or sociological information was gathered and stored on 5,011 separate individuals who met the criteria for inclusion.

In the period from 1951 to 1966 the population of Tokelau grew by 21% to 1,901, and then fell by 23% to 1,456 in 1982, increasing to 1,690 in 1986. One of the atolls (Nukunonu) showed a precipitous decline in numbers, with its 1982 population being 42% less than at its peak in 1966. Between 1982 and 1986 there was a considerable turnaround in the population of Tokelau and a large proportion of the losses from migration were made up by returning migrants.

Travel undertaken by Tokelauans proved to be an important consideration in the Tokelau Island Migrant Study. It became clear by the mid-seventies that travel experience had to be taken into account when assigning persons as nonmigrants or migrants. The extent of such travel was certainly greater than first anticipated. Three subgroups of islanders were defined.

1. The 'stereotypic' nonmigrant and 'stereotypic' migrant who had little or no travel experience.
2. Those with some travel experience: migrants who had left New Zealand for up to six months in Tokelau or elsewhere. Nonmigrants who had left Tokelau for up to six months in New Zealand or elsewhere.
3. Those with migration experience other than 'little' or 'some' were classified as having 'extensive' experience.

The effect of introducing detailed migration histories on the classification of Tokelauans as migrants and nonmigrants was examined in the data from a cohort of 689 adults aged 15-69 years at base line who were seen on Tokelau in Round I and followed through both the subsequent Rounds of the study in either Tokelau or New Zealand. The analysis showed that the 'stereotypic' nonmigrants accounted for only 40% of all those seen on the atolls at Rounds I, II and III, while 20% were considered to have had some travel and 40% an extensive amount of experience away from Tokelau during this period. Similarly only 39% of those migrants seen in New Zealand in Rounds I, II and III were 'stereotypic' who did not leave New Zealand after emigration. Twenty five percent had had some travel while 35% had been away from New Zealand extensively after immigration, often on return trips to Tokelau. Clearly the assumption that migration was a definitive one time experience could not be held. The Tokelauans have increasingly become a population moving freely back and forth between their home atoll and their adopted home in New Zealand. The way in which this movement influences the adaptation process, and the differences between those who do not travel and those who travel extensively, clearly have to be explored.

Diet in Tokelau and New Zealand

Variations in diet patterns among Tokelauans in Tokelau and New Zealand constitute an important part of the migration experience and can be linked with documented changes in weight.

Life in Tokelau had traditionally depended upon those few foodstuffs available in the atoll environment. Three tree-borne crops - coconut, breadfruit and *pandanus*, one root crop - *pulaka*, and a wide variety of fish comprised most of the native diet, and green coconuts (or coconut sap) provided a very large proportion of the fluid intake. In addition, pigs and chickens were raised (on the same basic foodstuffs as the humans), but eaten only at feasts or ceremonial occasions. With increased contact with, and cash from, the outside world, the availability of imported foods has increased.

Dietary studies, conducted on the atolls in 1968, 1971, 1976, and 1982, made clear the dominant position of the coconut as the atoll dwellers staff of life (Harding *et al.*, 1986). It supplied the largest part of their energy requirements, an estimated 69% in 1968, 62% in 1971, and 54% in both 1976 and 1982. Fish was the other major local source of food on Tokelau. Fishing is still a most important activity for the men but had changed dramatically by the early eighties with the complete replacement of traditional canoes by aluminium dinghies powered by outboard motors.

Between 1968 and 1976, the proportion of total energy supplied by imported foods doubled; this was particularly evident in the rapid increase in the consumption of sugars, which contributed only an estimated 2% of total energy in 1968 and 8% in 1976, increasing to 14% from sucrose in 1982. This is comparable to 13% in the quantitative diet study of Tokelauans in New Zealand in 1974/75. Gradually, therefore, imported foods are changing the nature of Tokelauan eating habits.

Migration to New Zealand brought immediate and far-reaching changes in the food-related behaviour of Tokelauans. Dairies and supermarkets took the place of the plantations and lagoons as sources of food, and money became absolutely essential. Coconuts, breadfruit and *pulaka* were either unavailable in New Zealand, or found only in a few stores catering to Pacific Islanders and at high prices. Fish was expensive, especially in relation to the cheaper cuts of meat. The result was that bread and potatoes supplanted the traditional vegetable foods as dietary staples, and meat became the prime source of animal protein. Tokelauans were slower to adopt other staples of the New Zealand diet such as eggs. There was, however, a tendency to broaden the range of foods used during the later seventies; this probably reflected both acculturative influences and increased affluence among many families (Harding *et al.*, 1986).

The result was a major change in both calorie intake and sources of nutrition for the migrants. The shift from fish to meats and dairy products led to a marked increase in cholesterol consumed by the migrants, and the overall mean energy intake per day increased by at least 350 calories. Because of the great reduction in coconut consumption, the migrants substantially reduced the proportion of their energy derived from fat at the same time as their cholesterol intake was rising.

The percentage of energy from different food groups has been documented, together with the changes that took place, over the period from 1968 to 1982. Over that period there was a fall in the energy contribution from coconut in Tokelau from 69% to 54%, compared with less than 3% in New Zealand; in New Zealand, meat, eggs and dairy products supply 50% of the energy compared to less than 5% in Tokelau. Fish provides a constant 10-12% of energy in Tokelau, dropping to 3% in New Zealand. This reflects the dramatic change in food available for consumption in New Zealand.

The pattern of diet changes and the increase in energy intake experienced by the migrants is clearly considerable. Diet must be regarded as one of the factors that have contributed to the changes in weight, health and disease patterns.

Weight and body mass change

By reference to the baseline weight data from a survey in Tokelau in 1963, those in Tokelau 1982 and in New Zealand 1980-81 (Figure 17.2) show a considerable increase in weight in most age groups over the period of 19 years, particularly in men and women in their middle years (Figures 17.2 and 17.3). The heaviest weights are seen in the migrants in 1980-1981; weight gain is clearly an important outcome of migration, but considerable weight gain has also occurred in those who remained in Tokelau.

The Body Mass Index, BMI (wt in kg / ht in cm^2), is now widely used as an estimate of body mass and obesity in population studies. In Western societies a body mass index of 3.0 units or above is considered as definitely obese and carries an increased mortality

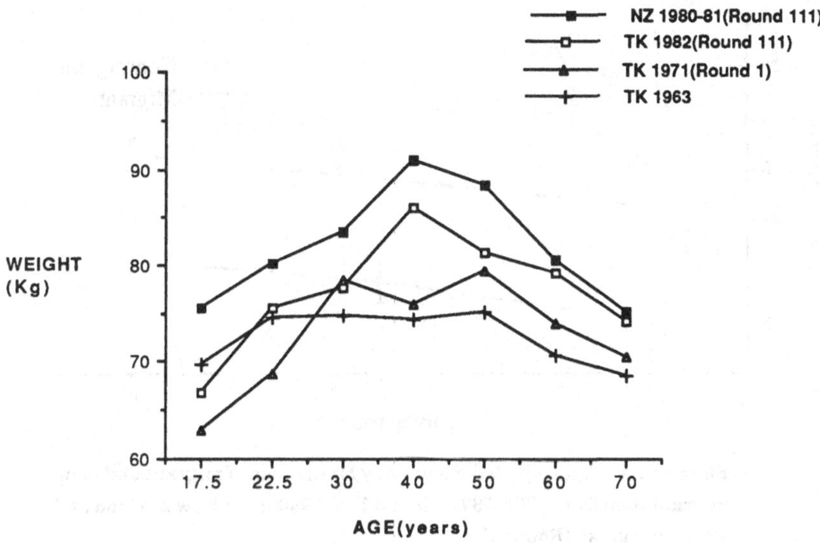

Figure 17.2. Mean age-specific weights in kg of men in Tokelau 1963 through to 1982 in Tokelau and to 1980-81 in New Zealand.

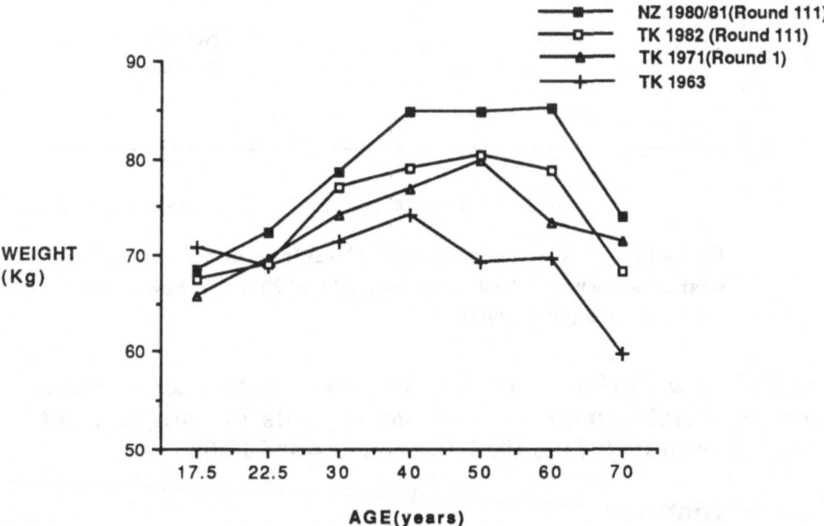

Figure 17.3. Mean age-specific weights in kg of women in Tokelau 1963 through to 1982 in Tokelau and to 1980-81 in New Zealand.

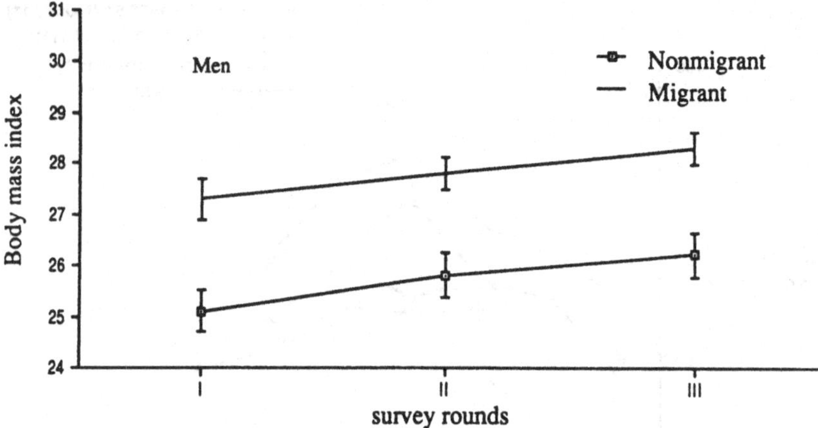

Figure 17.4. Age-adjusted mean Body Mass Index of migrant and non-migrant men from 1968-1971 (Round 1) to 1980-81 in New Zealand and 1982 in Tokelau (Round III).

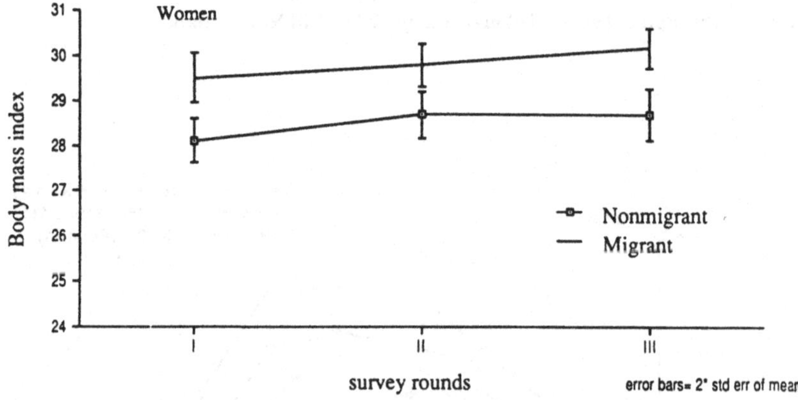

Figure 17.5. Age-adjusted mean Body Mass Index of migrants and non-migrant women from 1968-1971 (Round 1) to 1980-81 in New Zealand and 1982 in Tokelau (Round III).

risk (Lew & Garfinkel, 1979). The Body Mass Index increased among Tokelau migrants and nonmigrants in both men and women, from 1968-71 to 1982 (Figures 17.4 and 17.5).

Diabetes

Evidence is accumulating that rates of Type II or non-insulin dependent diabetes mellitus (NIDDM) increase with the changing diet and life styles associated with migration, urbanisation and modernisation (Prior & Davidson, 1966; Zimmet, 1982). This is true of the Tokelau population where increases in the prevalence and the incidence have occurred both in Tokelau and in the migrants in New Zealand between 1968 and 1982, and are greater in the latter, and especially in women (Ostbye *et al.*, 1989). In the men the prevalence between Round I and Round III increased from 30.2 per 1,000 to 69.7 per 1,000 in the nonmigrants and from 74.8 per 1,000 to 107.6 per 1,000 in the migrants. In the women, who have higher rates, the prevalence between Round I and Round III increased from 87.4 per 1,000 to 142.6 per 1,000 in the nonmigrants and from 117.2 per 1,000 to 198.5 per 1,000 in the migrants.

The age-standardised incidences over the period from 1968 to 1982 were also higher in the migrants than in the nonmigrants. Expressed as new cases per 10,000 person years at risk, the incidence was 124.6 and 253.8 in the migrant men and women respectively compared with 84.6 and 133.4 in the nonmigrant men and women. The age-standardised risk factors at entry in the Tokelauans show a number of important differences between those who did and those who did not go on to develop diabetes. These relate primarily to BMI, weight and systolic blood pressure. For example the mean weights in the men and women who did not develop diabetes were 79.12kg and 76.19kg compared with 87.73kg and 82.85kg in those who did.

These findings indicate the momentum of these biological changes in the migrants and give a warning of the public health problems that they pose to the individual, the communities and the area health boards in New Zealand. But also the incidence of diabetes in Tokelau is not low and is increasing. There is no doubt about the problem that effective management poses there now and will pose into the future.

Gout

Gout is a problem in a number of Polynesian groups (Prior *et al.*, 1966). In a long-term study of New Zealand Maori, a baseline prevalence of 8.8% and a high incidence of 10.3 per hundred over an eleven year period was reported in the men (Brauer & Prior, 1978). In the Tokelau study of hyperuricaemia and clinical gout the changes in diet, life style, work and living pattern in New Zealand over the active period of data collection (up to 14 years) allow the hypothesis to be tested that increased red meat and protein consumption, increased body weight, and a greater use of alcohol by some men place susceptible subjects with high uric acids at increased risk of clinical gout. The full results have been published (Prior *et al.*, 1987).

The place, age and sex-specific prevalences of gout in Round I and Round III indicate that it is more common in migrants than nonmigrants, and in men than women, and occurs at a younger age in the migrants. In Round I, 4 of the 10 cases were under 44 years while all the 9 nonmigrant cases were over 45 years. The condition became more common as migrants stayed longer in New Zealand. By Round III there were 16 cases out of a total of 33 that were under 45 years and no nonmigrant cases below that age. The metabolic changes needed to produce an increased uric acid pool and clinical episodes were clearly building up to a greater degree and at a faster tempo in the younger migrant men at risk. The age-standardised prevalence of gout in the migrant men increased from 21 per 1,000 in Round I to 51 per 1,000 in Round III while in the nonmigrants they fell from 19.5 per 1,000 in Round I to 14.6 per 1,000 in Round III. In Round III migrant men showed a significantly high relative risk of gout of 3.7.

Between 1968 and 1982 there were 1705 people included in the incidence survey of whom four nonmigrant men and 30 migrant men aged 35 and over developed gout, 17 of the migrants being in the 35-44 age group. Three nonmigrant women and five migrant women, all aged over 35, developed gout. Age-standardised incidences were 90.7 and 11.2 per 10,000 person years at risk in the migrant and nonmigrant men. The resulting age-standardised relative risk of developing gout for migrant versus the nonmigrant men was 9.0. Migrant women did not show a higher risk of gout than nonmigrant women. The men who developed gout were 8.4 years older than those who remained gout free. After controlling for age, those who developed gout had mean measures of body build, weight, skin folds and other variables including blood pressure, serum cholesterol, serum triglycerides and serum uric acids greater than the non-gouty men.

The difference in the incidence of gout between Tokelau and New Zealand may be related to several factors influencing purine metabolism associated with a build up of the total uric acid pool. These factors include the increased amount of purines ingested by the migrants with their high consumptions of red meat, such as beef, mutton and pork, the purine content of beer, and the influence which alcohol has to diminish renal uric acid clearance. In addition, a gain in body weight and increased muscle mass may also be contributing factors to the expanded uric acid pools in the migrants.

The prevention of weight gain, through moderate dietary changes, modest use of alcohol and an increase in physical activity may help to prevent gout in migrant Tokelauans.

Migration and blood pressure

In aiming to document the influence of migration on Western diseases in a population in whom their incidence was low prior to resettlement in a modern urban setting, the study considered blood pressure as an important indicator of the physiological effects of social change. Hypotheses were developed concerning the part played by physical and environmental factors compared to culture and social change.

Based on the 1968 examinations, the status of nonmigrants versus premigrant or future migrants was assessed by comparing 321 Tokelauan adults who were to migrate between 1968 and August 1973 with 592 who remained in Tokelau during that period (Prior *et al.*, 1974). While age and sex differences were factors in migration, with younger men dominating the vanguard, few physical differences were found between the future migrants and the nonmigrants suggestive of any biologically-determined selective biasses (Prior *et al.*, 1974).

The cross-sectional comparisons of blood pressures between the adult nonmigrants and the migrants show higher pressures in the migrants in almost all age groups. This pattern was found in each of the three Rounds. The data from Round I and from Round II have been published (Prior & Stanhope, 1980; Joseph *et al.*, 1983). The Round II results illustrate this pattern (Figures 17.6 and 17.7).

Among men, systolic and diastolic pressures were significantly higher among migrant than nonmigrants: after controlling for the effects of age and body mass, systolic pressures were 7.2mmHg higher and diastolic pressures 8.1mmHg higher. Thus, though differences in age structure and body mass were contributing to the observed difference, they explained relatively little of it. Among women, the relationship between migration and blood pressure was less dramatic; the increase in systolic pressure among migrants did not attain statistical significance and the mean adjusted diastolic pressures were 3mmHg higher among migrants. This gave clear evidence of a different pattern or different degree of response in the women compared to the men migrants.

Age and body mass have consistently been shown to be positively related to blood pressure in the different Rounds. The partial correlations between systolic blood pressure and body mass

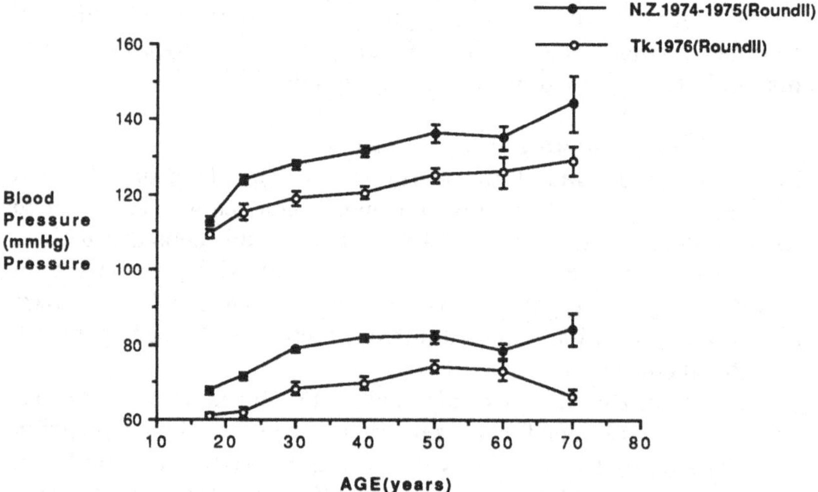

Figure 17.6. Age-specific mean systolic and diastolic blood pressures in migrant men in New Zealand 1974-1975 (Round II) and in non-migrant men in Tokelau 1976 (Round II).

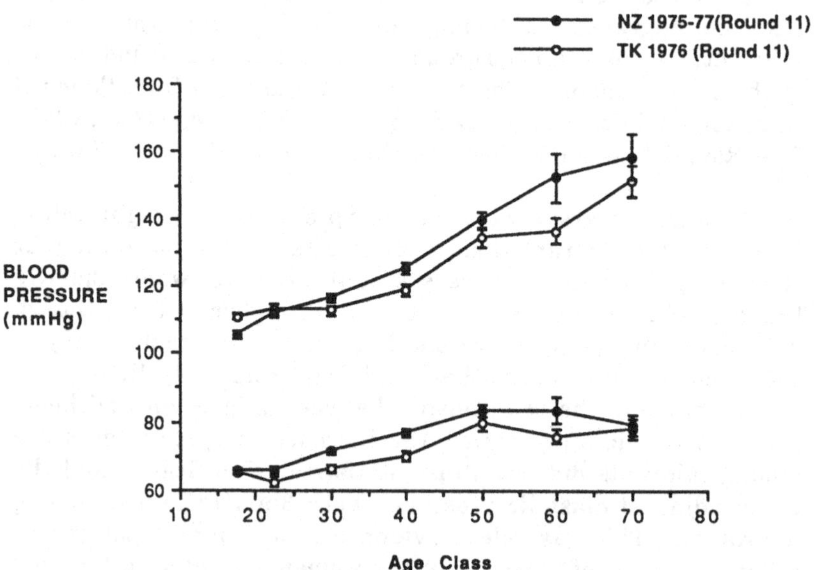

Figure 17.7. Age-specific mean systolic and diastolic blood pressures in migrant women in New Zealand 1974-1975 (Round II) and in non-migrant women in Tokelau 1976 (Round II).

(controlling for age) were 0.47 for men and 0.44 for women in Round II. There was no difference between migrants and nonmigrants.

Tokelau was identified as a low salt intake group from the time of the 1968 survey, with mean 24 hour sodium outputs of 40mmols increasing to 58mmols in 1982. Potassium outputs were in the 80-90mmols range over this period. In New Zealand the sodium intakes are notably higher based on early morning sodium concentrations of around 100mmols per litre and potassium of 50-55mmols per litre. There were no significant correlations between electrolyte measures and blood pressure in the early surveys. There is however evidence of a significant correlation of the sodium to potassium ratio with systolic pressure in the migrant men in the last two rounds of examinations, which suggests increased sensitivity to sodium developing with greater time in New Zealand.

The first opportunity to investigate possible migration effects longitudinally came when rates of blood pressure change from Round I to Round II could be analysed in the cohort of atoll-dwellers at Round I, some of whom had migrated by Round II (Salmond *et al.*, 1985). The hypothesis was that the rate would be higher in the migrants than the nonmigrants.

The analysed cohort of 812 individuals examined first in Tokelau in 1968 or 1971, and then in either Tokelau in 1976 or in New Zealand during the 1975-77 survey, consisted of 532 nonmigrants and 280 migrants. The mean length of follow-up was 6.2 years for migrants of both sexes and one year less for nonmigrants - 5.3 years for men and 5.2 years for women. Tokelauans examined in New Zealand had lived there for an average of 4.2 years if male, and 4.1 years if female.

Rate of change measures were defined as the difference between follow-up and baseline values divided by the number of years between the two examinations. The only variables which were shown to have any statistically significant relationship with rate of change in blood pressure were the baseline variables - age, body mass index and blood pressure - and also the rate of change in body mass. Age-specific comparisons of the rates of change in systolic and diastolic pressure showed that the raw blood pressures of migrants were increasing, while those of nonmigrants were decreasing slightly. Striking differences were found in the men, with mean systolic pressures 10mmHg higher among the migrants over the five to seven years of follow-up.

Sex-specific analyses of covariance were undertaken to provide more accurate estimates of the relationship between migration and rate of change in blood pressure. After adjustment for all relevant covariates including age and body mass index, systolic and diastolic blood pressures of male migrants were each found to have risen 1mmHg per year faster, on average, than in male nonmigrants. Less than half this difference was observed in women, 0.41mmHg for systolic pressure and 0.42mmHg, for diastolic pressure. The important effect of migration was clearly influencing women less than men.

The longitudinal relation between blood pressure and migration was further examined in a cohort of 654 adult Tokelauans through three survey periods from 1968 and 1982.

These results confirmed that both men and women increased their body mass on migration to a greater extent than the nonmigrants. Both systolic and diastolic blood pressures of migrant men were significantly higher than would be expected in this cohort on the basis of age when compared to the nonmigrant men. The differences in the women were smaller.

The study of the overall migration component of the effects observed (of the 'non body mass component', and also the travel experience effect) compared the baseline examinations of 1,018 adult Tokelauans aged 15-19 years in Tokelau either in 1968 or 1971 (Salmond *et al.*, 1989) with those members of the original cohort who were also examined at Round II and Round III. After exclusions there were longitudinal records for 654 subjects available for analysis. These included 251 who had had 'little' travel experience who were the principal group studied in order to focus on the purest migrant and nonmigrant effects. Since the data consisted of three records per person, regression models involving generalised least squares estimation methods were used. One analysis estimated the overall migration effects and the other the 'non body mass effects', those unexplained by body mass.

By Round II, the increases in the mean annual rate of change in systolic and diastolic pressures adjusted for age, body mass, and initial blood pressure of the migrant over the nonmigrant men were 1.05 and 1.22mmHg per year respectively. These figures compare with 1.03 and 1.22mmHg respectively in the earlier study. Among migrant women, by Round II, their rate of change in systolic pressure per year was 0.63mmHg and 0.41mmHg for diastolic pressure. The estimates from the regression models of the mean adjusted overall migration effect and the 'non body mass' components of systolic and diastolic pressures are shown graphically in Figure 17.8 for men and Figure 17.9 for women.

By Round II the overall period migration effects for systolic and diastolic mean pressures in men (Figure 17.8) were larger

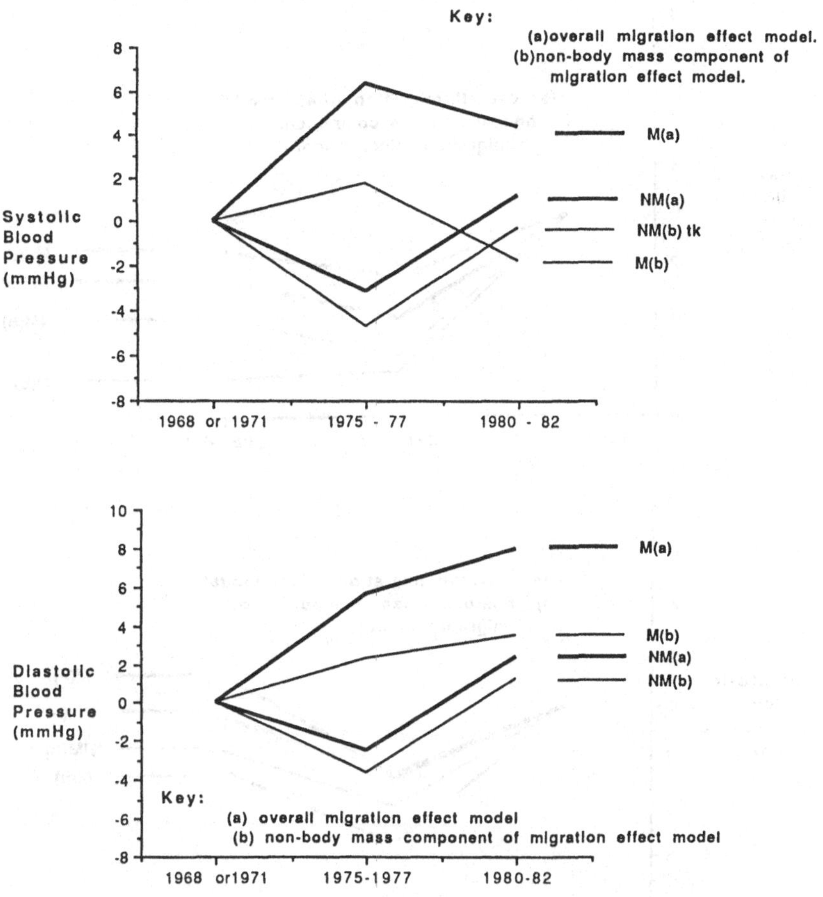

Figure 17.8. The generalised least squares regression models for estimating the mean adjusted 'overall migration' effect and the 'non-body mass' components of the systolic and diastolic pressures for men from Round I to Round III.

than the non body mass components at 9.50mmHg and 8.19mmHg respectively above the baseline level in 1968 or 1979. Part of this difference is due to a fall of 3.16mmHg in Tokelau pressures. The mean adjusted 'non body mass' systolic and diastolic pressures among the migrants were 6.42mmHg and 5.86mmHg higher than those among the nonmigrants. The pattern in the women (Figure 17.9) is very different from that in the men and the migration effects are negligible for both systolic and diastolic pressures. At Round III there is less overall migration effect for either men or women, with an adjusted mean difference for the men of 3.2mmHg

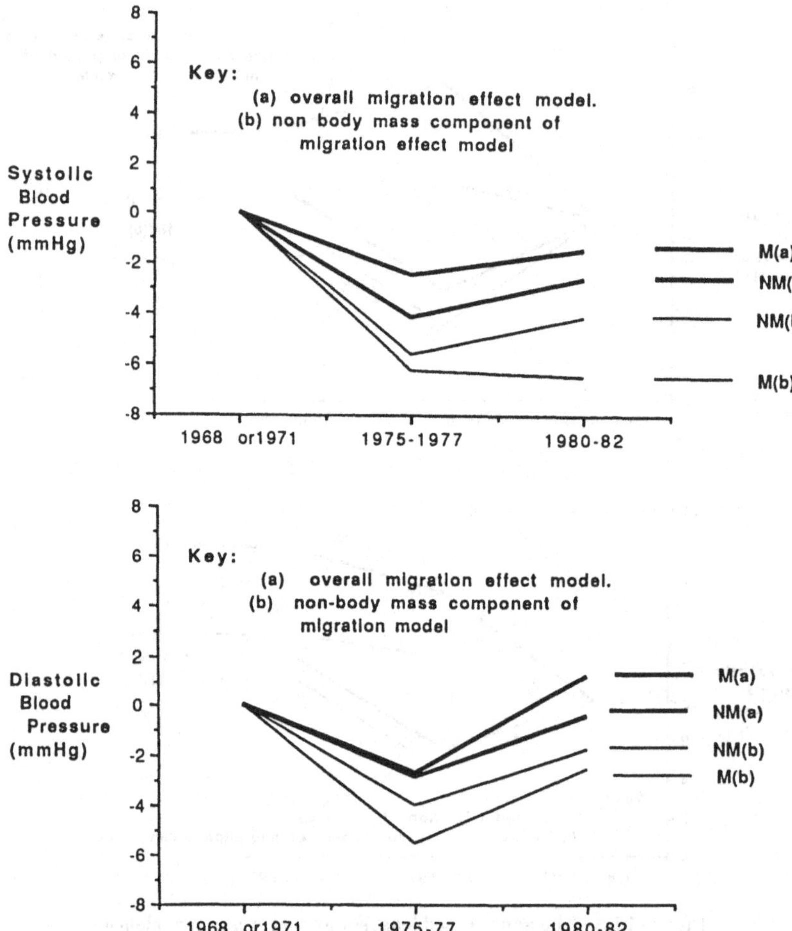

Figure 17.9. The generalised least squares regression models for estimating the mean adjusted 'overall migration' effect and the 'non-body mass' components of the systolic and diastolic pressures for women from Round I to Round III.

systolic and of 5.6mmHg diastolic pressures; the latter includes a small 'non body mass' contribution of 3.5mmHg in the migrants. It is possible that environmental and social changes contributing to diastolic pressure changes take longer to evolve. By Round III the main part of the overall migration effects can be explained by body mass changes as the 'non body mass' contributions have diminished. Some decline of mean adjusted systolic and diastolic pressures in women can be seen in both migrants and nonmigrants.

What stands out from these analyses is the difference in response over time of the men and the women. The men show a significant increase in 'overall migration' effect but lose any 'non body mass' component by Round III. The data for the women show a decrease from baseline in both the migrants in New Zealand and those in Tokelau in strong contrast to the findings in the men. The possibility of apressor response in the 1968-71 community has to be considered, but if it occurred it appears to have affected the women more than the men. The question arises as to whether the women were in some way protected from factors in the New Zealand environment that might be expected to lead to increased blood pressure, in contrast to the men where blood pressures rose and stayed elevated. The variation with travel also had a smaller effect in women than men and may relate to some of the biological effects of migration and location changes.

Over the period of the study the 'extensive travel' group experienced less increase in Body Mass Index than the other two groups. This may be one of the factors whereby return visits to Tokelau contribute to lower blood pressure in some subjects. An unexpected finding is that blood pressures have not been stable in Tokelau over the study period, and this is shown in particular by the drop observed in men and women between the early and mid-1970s. Factors possibly contributing to these changes on a population basis have been reported (Salmond *et al.*, 1989).

Data have also been collected to examine the role of psychosocial factors, social networks and coping strategies in relation to blood pressure in the migrants. Those migrants interacting more with non-Tokelauans and moving more into the New Zealand community had higher blood pressures than those with a mainly Tokelau network. The contribution to the total variance of blood pressure is small, with body mass contributing more than five times that of the social interaction variable (Beaglehole *et al.*, 1977).

Indirect evidence of stress was found in studies of 24 hour urine output of catecholamines where the nor-adrenaline levels were notably higher in Tokelauans in New Zealand than in men in Tokelau. Those from Tokelau were also lower than those in samples from the United Kingdom, the United States, Japan and Nigeria (Jenner *et al.*, 1987).

In their different ways these results indicate that areas of conflict and stress are influencing the blood pressures of Tokelauan men in both Tokelau and New Zealand but they contribute less than weight gain.

The Tokelau pedigree file has been used to examine the

interaction between the genetic and cultural environmental factors in migrant and nonmigrant Tokelauans (Ward *et al.*, 1980). The familial aggregation of blood pressure in first degree relatives in Tokelauans is essentially similar to that reported in a wide range of different ethnic groups (Ward, 1990). But estimates of heritability for systolic pressure rose from 21.9% in the nonmigrants to 34.4% in the migrants while the cultural transmission factors remained stable, 22% and 25.6% respectively. This raises the question as to whether the increase in the heritability in migrants may be due to a subpopulation of genetically predisposed 'responders' to the new environment.

Conclusions

The data indicate that a number of changes have occurred in the health and disease status of Tokelauans in 1968-1982. Body weight, obesity, blood pressure, and risk of conditions such as diabetes and gout are notably increased in the migrants in New Zealand. The increase in blood pressure is more marked in the men than the women.

Migration, modernisation stress, and culture changes affect men more than women. The reasons for the apparent sparing of women from this process may lie in social and cultural supportive factors with the Tokelau family in New Zealand. There is some evidence from the Tokelau study that stress, including the dissonance between status and privately held views, can influence blood pressure. There is an apparent beneficial effect of return visits to Tokelau on blood pressure and weight, perhaps through a return to a Tokelauan diet and an opportunity for cultural renewal and a holiday. Changes are occurring in Tokelau with increased availability of Western style food, a move into a cash economy, the reorganisation of the society and the status of the people within it. However there is no evidence of a general effect on blood pressures which were in fact lower in 1982 then when the study commenced in 1968. At the same time weights are increasing and diabetes is becoming more of a problem in Tokelau. In both Tokelau and New Zealand these problems will need to be faced by innovative methods of health education and health promotion led by the communities themselves.

Acknowledgements

The author acknowledges the support of this project given by the Medical Research Council of New Zealand, the Cardiovascular Unit of the World Health Organisation, The Wellington Hospital

Board and in particular the Tokelau people in Tokelau and New Zealand.

References

Beaglehole, R., Salmond, C. E., Hooper, A., Huntsman, J., Stanhope, J., Cassel, J. C. & Prior, I. A. M. (1977). Blood pressure and social interaction in Tokelauan migrants in New Zealand. *Journal of Chronic Diseases*, **30**, 803–12.

Brauer, G. W. & and Prior, I. A. M. (1978). A prospective study of gout in New Zealand Maoris. *Annals of Rheumatic Diseases*, **37**, 466–72.

Cassel, J. C. (1974). Comment: hypertension and cardiovascular disease in migrants: a potential source of clues? *International Journal of Epidemiology*, **3**, 204–6.

Elliott, R. A. (1963). Tokelau health mission: report to department of Maori and island affairs. Government of New Zealand, Wellington.

Goodenough, W. H. (1963). *Cooperation in change: an anthropological approach to community development*. New York: Russell Sage Foundation.

Harding, W. R., Russell, C. E., Davidson, F. & Prior, I. A. M. (1986). Dietary surveys from the Tokelau Island Migrant Study. *Ecology of Food and Nutrition*, **19**, 83–97.

Jenner, D. A., Harrison, G. A. & Prior, I. A. M. (1987). Catecholamine excretion in Tokelauans living in three different environments. *Human Biology*, **59** (1), 165–172.

Joseph, J. G., Prior, I. A. M., Salmond, C. E. & Stanley, D. (1983). Elevation of systolic and diastolic blood pressure associated with migration: the Tokelau Island migrant study. *Journal of Chronic Diseases*, **36**, 507–16.

Lew, E. A. & Garfinkel, L. (1979). Variations in mortality by weight among 750,000 men and women. *Journal of Chronic Diseases*, **32**, 563–76.

Ostbye, T., Welby, T. J., Prior, I. A. M., Salmond, C. E. & Stokes, Y. M. (1989). Type 2 (non-insulin-dependent) diabetes mellitus, migration and westernisation: the Tokelau Island migrant study. *Diabetologia*, **32**, 590–595.

Prior, I. A. M. (1974). Cardiovascular epidemiology in New Zealand and the Pacific, *New Zealand Medical Journal*, **80**, 245–252.

Prior, I. A. M. & Davidson, F. (1966). The epidemiology of diabetes in Polynesians and Europeans in New Zealand and the Pacific. *New Zealand Medical Journal*, **65**, 375–83.

Prior, I. A. M., Rose, B. S., Harvey, H. P. B. & Davidson, F. (1966). Hyperuricaemia, gout and diabetic abnormality in Polynesian people. *Lancet*, **1**, 333–8.

Prior, I. A. M. & Stanhope, J. M. (1980). Blood pressure patterns, salt use and migration in the Pacific. In *Epidemiology of arterial blood pressure: developments in cardiovascular medicine*, ed. H. Kesteloot & J. V., Joosens, pp. 243–62. The Hague: Martinus Nijhoff Publishers.

Prior, I. A. M., Stanhope, J. M., Evans, J. G. & Salmond, C. E. (1974). The Tokelau Island migrant study. *International Journal of Epidemiology*, **3**, 225–32.

Piror, I. A. M., Welby, T. J., Ostbye, T., Salmond, C. E. & Stokes, Y. M. (1987). Migration and gout: the Tokelau Island migrant study. *British Medical Journal*, **295**, 457–61.

Salmond, C. E., Joseph, J. G., Prior, I. A. M., Stanley, D. G. & Wessen, A. F. (1985). Longitudinal analysis of the relationship between blood pressure and migration: The Tokelau Island migrant study. *American Journal of Epidemiology*, **122**, 292–301.

Salmond, C. E., Prior, I. A. M. & Wessen, A. F. (1989). Blood pressure patterns and migration: a 14-year cohort study of adult Tokelauans. *American Journal of Epidemiology*, **130** (1), 37–52.

Ward, R. (1990). Familial aggregation and genetic epidemiology of blood pressure. In *Hypertension, pathophysiology diagnosis and management*, ed. J. H. Laragh & B. M., Brenner. pp. 81–100. New York: Raven Press Ltd.

Ward, R. H., Raspe P. D., Ramirez, M. E., Kirk, R. L. & Prior, I. A. M. (1980). Genetic structure and epidemiology: the Tokelau study. In *Population structure and genetic disorders* ed. A. W., Eriksson, pp. 301–325. London: Academic Press.

Zimmet, P. (1982). Type 2(non-insulin-dependent) diabetes: an epidemiological overview. *Diabetologia*, **22**, 399–411.

18 *Micromigrations of isolated Tuareg tribes of the Sahara Desert*

PHILIPPE LEFEVRE-WITIER

If the main objective of anthropology is the study of variation among and within human groups (Gomila, 1976), it is necessary to 'determine first the structure and distribution of any identifiable units' (Roberts, 1965). But this task presents difficulties. Such units are fluid. Human populations are groups of living and reproducing entities. Like any living system, they demonstrate a combination of different levels of organisation and integration, primary as cells, secondary as tissues and, yet more complex, tertiary as the whole morphology. Obviously these different levels and their combinations have different functions, the regulation and evolutionary meaning of which have barely been explored.

It is convenient to distinguish two different levels of population organisation and structure (Lefevre-Witier, 1976). A first level of discontinuity is that of the 'genetic population'. This consists of the mating circle and the circle of fertile offspring. This definition differs from that of isolates of Dahlberg (1948) which concerns mating only; from that of the panmictic mendelian populations of Dobzhansky (1970) which extends the circle up to panmixia but excludes mate selection; and from the isogamic population of Malecot (1966) which also excludes selection. These three have in common a human group where the maximum of gametic transmission and exchange occurs as well as a strong tendency to genetic homogeneity; the principal limitations are the absence of differential fertility and infant mortality generation by generation expressing the action of natural selection. Our definition is more comparable to the endogamic circle of Henry (1968) and the natural effective population of Wright (1946). The durability of such genetic populations is dependent upon their genotypic and panmictic structure, varying to some extent with the demographic size of the genetic population. It also depends on neighbouring populations and the number of mate exchanges that occur with them, and on environmental conditions.

A second level of discontinuity depends on the concept of 'limited size isolated populations', their isolation being imposed by barriers of many different types - territorial, climatic, linguistic, social, technological. These barriers enforce strong links between small populations and their conditions of life, their environment, and sometimes restrict their existence to a very sharply demarcated ecosystem. The concept however also includes situations where such small populations form part of a larger human stock which can be considered as more continuous. Many of these barriers are under the control of man himself and thus provide variable evolutionary conditions with different evolutionary results. A good example of the genetic consequences of history and territorial politics is the recent demonstration by Sokal *et al.* (1989) of the high correlations between discontinuities in gene pools and linguistic frontiers. But the study of many small populations shows that the two types of discontinuity, genetic and ecological, do not always coincide. Different genetic populations can develop within the same set of barriers or within the same ecosystem. Similar gene pools can be modified by any barriers that impose different conditions of selection or mate exchange.

Study of the discrepancies between geno- and eco-populations is of great importance. Selecting the pertinent discontinuities in populations is equivalent to selecting a focal distance or a power of resolution in any laboratory technical procedure. Scientific reality is related to the tool that is used to observed it.

The Tuareg

The Tuareg groups of the central Sahara Desert, and the genetic exchanges that occur within and among them, provide good examples. Until recently, very little was known of their biological and anthropological structures, and such information as there was was mostly collected by observers in the French army from 1902 onwards. There are many small groups, their small size being associated with the very severe environmental conditions. Many years ago Capot-Rey (1953) used the term 'molecular humanity' to describe these Sahara groups. The Tuareg kinship system is both patrilineal and matrilineal, and thus provides the unusual possibility of reconstructing complete family trees.

Three examples of field work in the Tuareg area (Figure 18.1) are presented to illustrate some of the problems of population definition: (a) the population structures of eleven small nomadic Tuareg groups living under similar environmental conditions in the Tassili n'Ajjer mountains (Algeria); (b) the structure and long-term genetic evolution of two Iwellemeden Tuareg tribes under

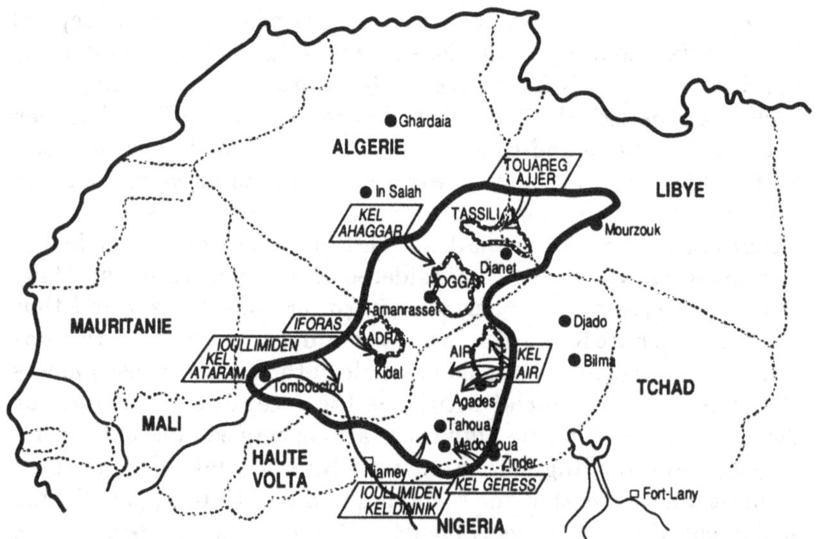

Figure 18.1. Geographical location of the three Tuareg tribal groups studied.

different social conditions in Mali and Niger; (c) the recent genetic evolution of sedentary groups in Ideles, a village in the Tuareg area of the Ahaggar mountains (Algeria).

Tuareg Isseqqamaren

Between 1970 and 1980 our team undertook researches among the northern confederation of Tuareg tribes of the Algerian Tassili n'Ajjer and Ahaggar tropical mountains. The tropic of Cancer passes near Tamanrasset, the main town of the Ahaggar region.

The genetic constitution of the small groups of 30 to 100 Tuareg of the Tassili n'Ajjer (Figure 18.2), each apparently autonomous, was studied using the serogenetic characters of their red blood cells and serum. The first study revealed, for each group, quite strict endogamy and precise tribal names according to their pastoral territories located in the large but practically identical environments of arid plateaus and wadis (Kel Tefedest, Kel Amguid, Kel Intounin, Kel Ohet, etc.).

The Gm system, with its characteristic haplotypes, is perhaps the most informative in this area (Table 18.1). It demonstrates (1) considerable homogeneity for each group, with a gene-pool limited to three, four or five haplotypes, (2) strong differentiation from group to group, with very specific distributions of frequencies, and (3) departure from Hardy Weinberg equilibrium in each group.

Table 18.1. *Gm haplotype distribution in some Tassili n'Ajjer Tuareg groups*

Haplotype	*Iheyawen Hada N=25*	*Kel Inghar N=43*	*Kel Amguid N=66*	*Kel Tefedest N=67*
Gm 1,17,21	.1898	.3452	.1666	.5176
Gm 4,5,10,11,13,14	.1104	.2776	.6668	.3148
Gm 1,17,10,11,13,15,28	.4216	.1681		
Gm 1,17,5,14	.0828			
Gm 1,17,5,10,11,13,14,28		.1946		
Gm 1,17,5,10,11,13,14			.1488	

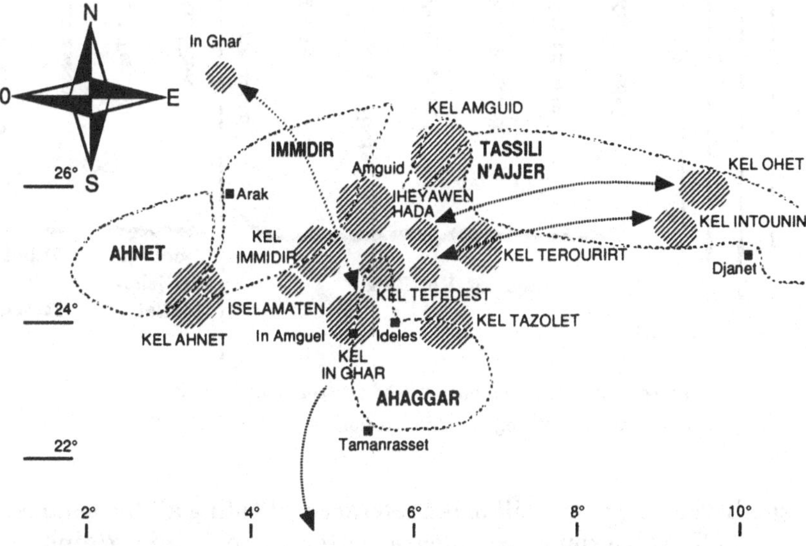

Figure 18.2. Tassili n'Ajjer Isseqqamaren Tuareg nomadic groups.

The question was whether these groups were isolates far from panmixia, and (some of them) so limited in their demographic capacity to reproduce that they were condemned to disappear, or did they form part of a larger gene-pool that suffered severe segmentation and isolation by the conditions of the environment? From several different studies these questions were answered. First, historical research revealed the existence, a long time ago, of a group of migrants called Isseqqamaren, to which most of the

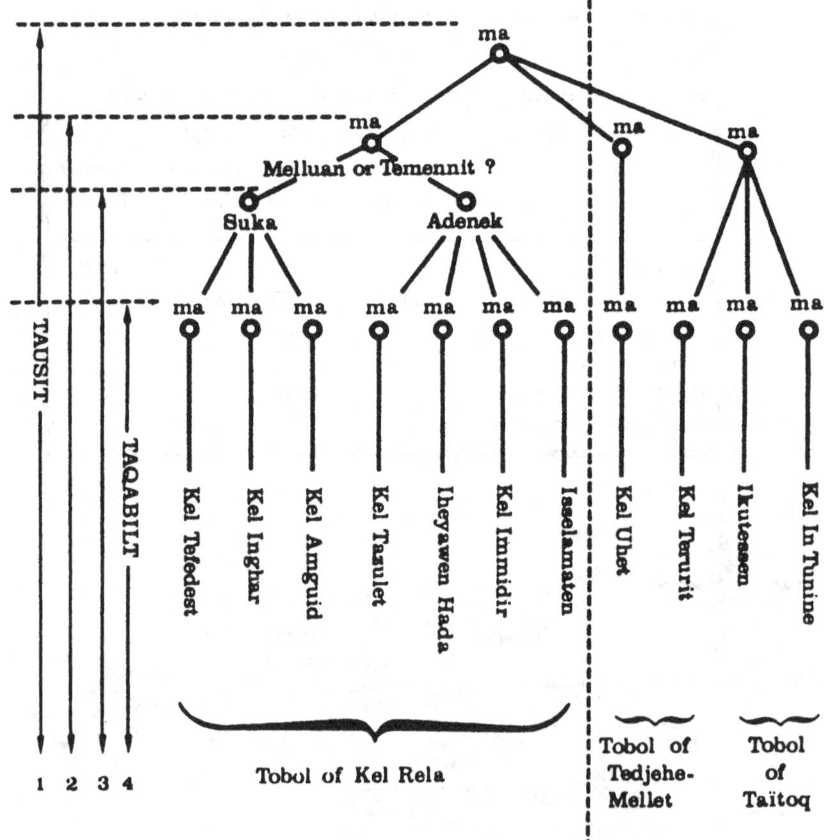

Figure 18.3. Maternal parenthood and segmentation for 11 Isseqqamaren Tuareg nomadic groups.

small Tassili groups still make reference. Pooling all the samples of the Tassili groups provided a better fit to Hardy Weinberg equilibrium of genotype and gene frequencies, similar to that of other larger Tuareg tribes. Tracing the pedigrees of the groups for 10 generations demonstrated links through the maternal lines since the earliest times, as well as very limited exchange of wives among some of the eleven groups at each generation (Figure 18.3). Migration of only a few individuals from one small group to another is capable of producing a panmictic effect (Jacquard, 1974). The conclusion was that the migration of Isseqqamaren, followed by their peopling of the Tassili n'Ajjer, was accepted as an historical and demographic fact, and the 10 to 12 ecogroups are the outcome of fragmentation of the original migrant population and

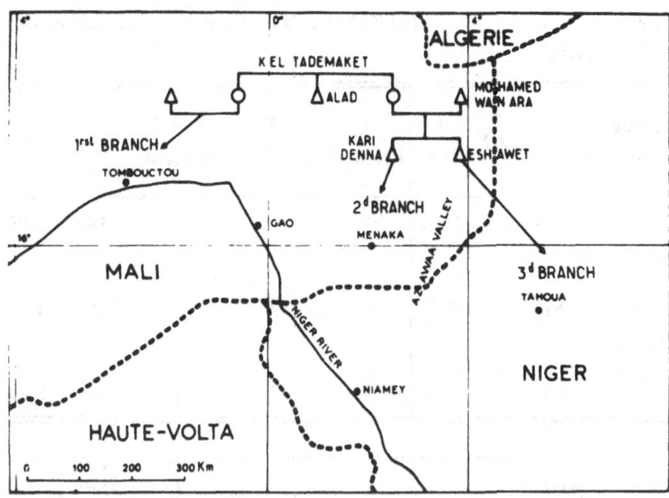

Figure 18.4. Geographical segmentation and distribution in the Sahel of the Tuareg descending from the Kel Taddemaket, the Kel Kummer (Kari Denna), and Kel Nan (Eshawet) (Chaventre, 1972).

their descendants due to ecological pressures, the whole making up the genetic entity of the Isseqqamaren Tuareg.

Tuareg Kel Kummer

The southern Tuareg confederation of Taddemaket or Iwellemeden tribes provides a second example (Figure 18.4). Here the results of field work in 1973 and 1974 are quite different from the Isseqqamaren. For social and political reasons a strict genetic isolate was generated from a large nomadic group intramarrying probably for centuries before the 17th. This isolate is known as the Kel Kummer. Their closest cousins are the Kel Nan and they separated from each other 300 years ago for political reasons. Besides our investigations in the CNRS laboratory in Toulouse, both tribes were studied by our colleagues in France, A. Chaventré, A. Jacquard and L. Degos (Lefevre-Witier & Jacquard, 1974).

The HLA typing at locus A and B (carried out at Saint Louis Hospital, Paris), demonstrated strong drift and founder effect, and so do other markers. There are specific distributions of the HLA haplotypes (Figure 18.5) and of the variant markers such as the abnormal haemoglobin D Ouled Rabah (Table 18.2) and an African glucose-6-phosphate-dehydrogenase A variant. The HLA haplotype distributions, combined with pedigree studies over 16 generations, show the paths of descent of the main haplotypes

Table 18.2. *Haemoglobin D Ouled Rabah and thalassaemia in Iwellemeden Tuareg (Niger and Mali)*

Tuareg Iwellemeden	*AD*	*DD*	*% D+*	*A_2 Elevated*	*F Elevated*	*A_2+F present*	*Total*	*%*
Kel Nan (Niger) 154	8	-	5.19	7	3	4	14	9.09
Kel Kummer (Mali) 289	57	3	20.76	-	3	-	3	1.04
Total	65	3	15.35	7	6	4	17	3.84

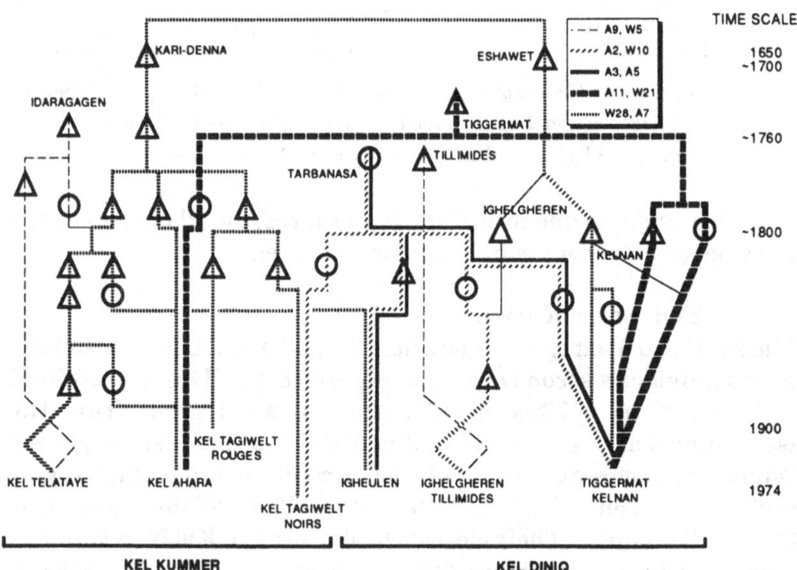

Figure 18.5. Family transmission and tribal distribution of some HLA haplotypes in the Taddemaket and Iwellemeden groups.

encountered in both groups, as well as the way that some of them disappear from each gene-pool. The specific content and homogeneity of the gene-pools, the strictly endogamous family trees, and high probabilities of gene origin from the Kel Kummer founders (today still 80% of genes in generation 16 come from the first 15 founders, Table 18.3) prove that the Kel Kummer is a true genetic isolate.

In conclusion the two Tuareg groups, Kel Kummer and Kel Nan, developed such physical and structural distance between

Table 18.3. ***Probability of origin ($\times 10^{-3}$) of genes by generation and after 16 generations in the Kel Kummer Tuareg isolate*** (Jacquard 1972)

Generations	*1-5*	*6-7*	*8-9*	*10-11*	*12-13*	*14-16*	*All gens*
No. of individuals	*126*	*371*	*601*	*653*	*268*	*245*	*2264*
Founders							
1093 & 1096	198	142	158	146	155	186	157
1 &2	199} 571	110} 350	226} 370	105} 357	105} 361	116} 426	115} 380
1331 & 1332	174	98	106	106	101	124	108
1919	30	66	76	64	71	107	71
1959	47	55	70	60	63	71	63
2060	16	60	48	33	37	33	41
2067	16	21	49	45	43	46	40
1968	29	35	39	34	40	53	38
2062	0	44	25	24	25	27	27
1628 & 1629	38	20	19	15	19	24	20
2063	0	23	23	15	22	20	19
15 main founders	747	674	729	647	681	807	699
10 following founders	103	112	102	111	101	78	104
131 later founders	150	214	169	242	218	115	197

them that their traditional mating system and circle disintegrated from generation to generation, leading to the two populations who are today very different. The areas that they inhabit are now widely separated. The Kel Kummer are settled in Mali near Gao, and the Kel Nan in Niger near Tahoua. They differ in their gene-pools and genetic structure. They also differ in their recent fate. In May 1990, 1,700 Tuareg were killed near Tahoua by the army of Niger and since then there has been no news of the fate of the Kel Nan families.

Ideles

This village in the Tuareg area of the Ahaggar mountains provides another different example in its genetic structure and evolution, as shown by the study of its pedigree structure and genetic constitution.

Various individuals and family groups from the Sahara came to Ideles 150 years ago, at the invitation of the leading Ahaggar

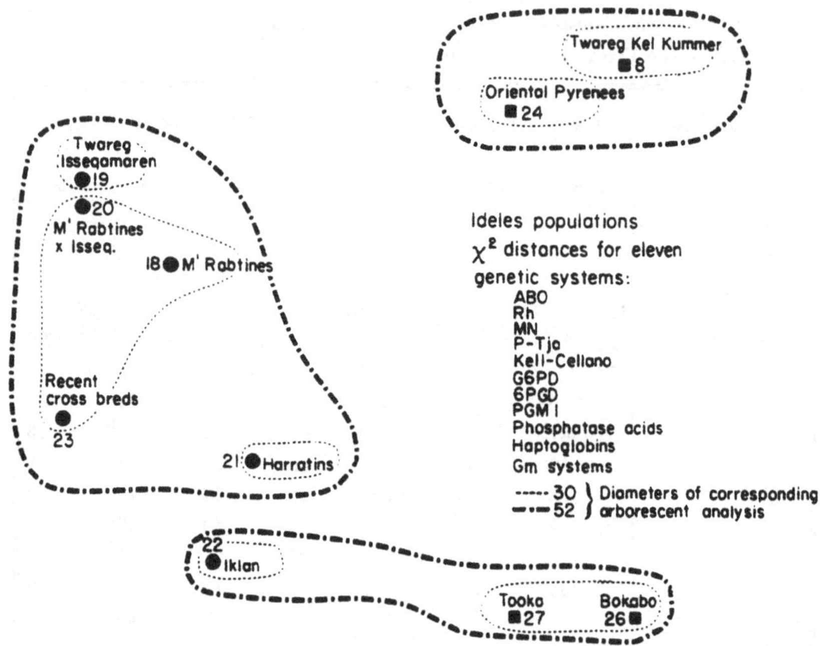

Figure 18.6. Principal component analysis of the genetic markers in the six main groups of Ideles inhabitants (Red cells: ABO, Rheus, P-Tja, MNSs, Kell-Cellano, Kidd. Serum: Gm, Inv, Hp and Gc).

Tuareg tribe. They were offered as much land and water as they could obtain by their work. A number of hamlets were built, the distances between them giving rise to or substantiating pre-existing barriers between the subgroups. Such other sources of differentiation were their geographical origins, their physical types ranging from Maghreb Mediterranean to black African, religions ranging from strict Islam to Animism, and their social structure, following the Tuareg hierarchy, ranging from independent individuals to domestic slaves.

The relative isolation of Ideles, the severe ecological conditions that make difficult food production and survival, and the fertility differentials, led the groups to develop into an integrated community. Genetic hybridisation slowly increased from generation 1 to generation 6/7 in 1970. The family trees of the 500 present day inhabitants were constructed.

In this population, the probability of origin of genes is more a measure of inter-group movement than intra-group, but it also indicates the main lines of gene flow within Ideles. In each group a founder effect can still be discerned, and the situation is far from

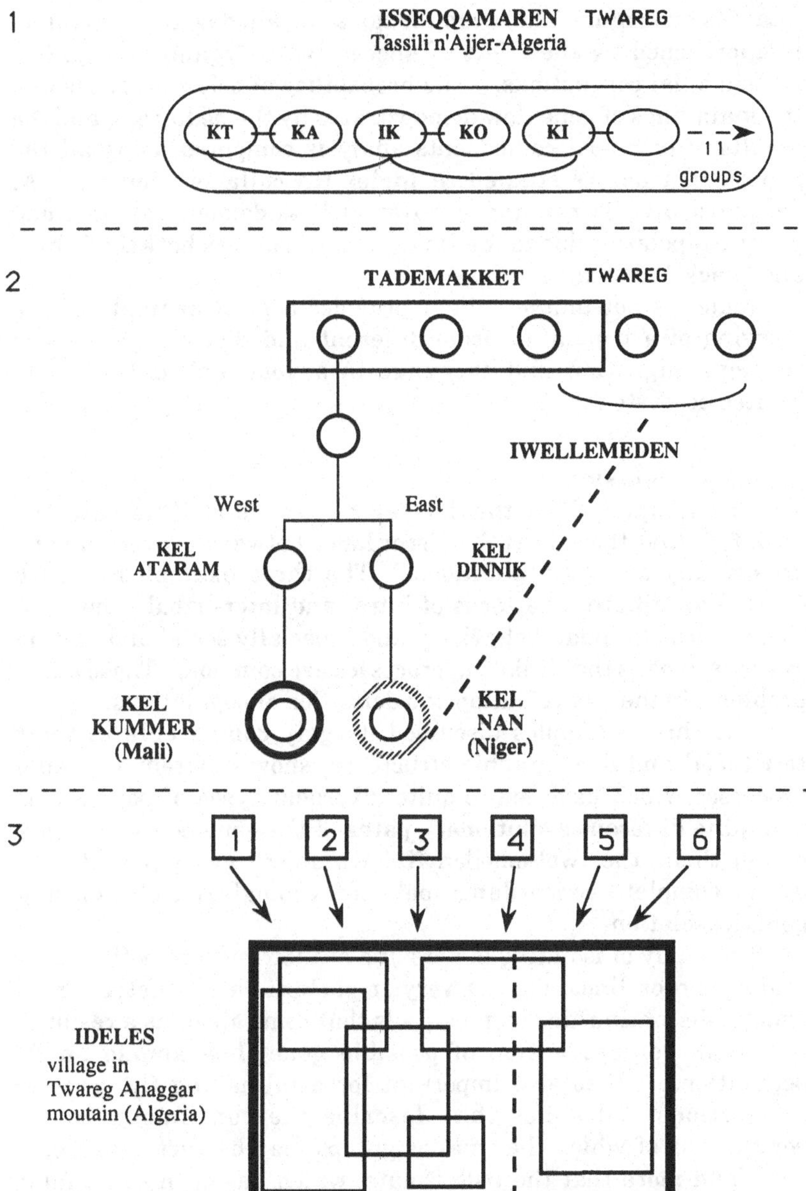

Figure 18.7. Schematic social and genetical structures of the three Tuareg populations presented as field work examples.

total fusion (Figure 18.6) even though a certain degree of panmixia is approached by the effect of successive polygamy and extra-matrimonial conceptions. The probability of origin of genes and the main lines of gene flow demonstrated by the pedigrees, and the results of principal components analysis computed using all the genetic characters studied in Ideles (in collaboration with A. Jacquard, J.L. Serres, and D. Salmon), both demonstrate a strong genetic bipolarity due to the strong social barriers between 'white' and 'black' Saharans.

Ideles is not unique, but it provides a good example of the merging of a population from different and distinct gene-pools through migration and exchange in a very well-defined and limited ecosystem.

Conclusion

These examples from the Tuareg populations illustrate the variation that there is in the discordance between genetically and ecologically defined populations. The three examples (Figure 18.7) show different patterns of intra- and inter-tribal gene flow. The nature of human behaviour, and especially social and mating systems, means that isolating processes are common. These cause problems in the identification and definition of populations.

The three examples discussed, largely comparable in their territorial and demographic structures, show different isolating processes which have led to quite divergent types of populations and quite different evolutionary paths of the gene-pools. In only one of them, the Iwellemeden Kel Kummer, was it possible to accept complete concordance between ecological isolation and genetic isolation.

The study of isolating factors and of their hierarchy (barriers, limits, border lines, etc.) is very important in the detection of genetic discontinuities in gene-pools that come about as a result of restriction or prevention of possible gene flow among small populations. It is also important for establishing the genetic correlations and clines that describe the continuous larger populations of which the smaller groups may be part. It seems more and more that the true isolate, which has been so useful a tool in mathematical and statistical models, may be less useful in our present understanding of population dynamics and evolution.

References

Capot-Rey, R. (1953). *Le Sahara Français*. Paris: Presses Universitaires de France.

Chaventre, A. (1972). *Les Kel Kummer, un isolat du Sud-Sahara. Population*. No. 4–5. Paris: INED.

Dahlberg, G. (1946). *Mathematical methods for population genetics*. Bale and New York: S. Krager.
Dobhzansky, T. (1970). *Genetics of the evolutionary process*. New York: Columbia University Press.
Gomila, J. (1976). Definir la population. In *L'étude des isolates*. pp. 5–36. Paris: INED.
Henry, L. (1968). Problemes de nuptialite – considerations de methodes. *Population*. **23**, 835–844.
Jacquard, A. (1972). *Evolution du patrimoine génétique des Kel Kummer Population*. No. 4–5. Paris: INED.
Jacquard, A. (1974). *The genetic structure of populations*. Springer-Verlag.
Lefevre-Witier, Ph. (1976). Populations génétiques at populations 'isolatées': definition, exemples, remarqués. In *L'etudes des isolates*. pp. 43–47. Paris: Espoirs et limites.
Lefevre-Witier, Ph. & Jacquard, A. (1974). Un 'isolate' du Sud-Sahara: les kel Kummer VI: structures génétiques des systèmes sanguins erythrocytaires et sériques VII. Conclusions provisoires. **3**, 518–534. Paris: INED.
Malecot, G. (1966). *Probabilites et Heredité*. Paris: INED.
Roberts, D. F. (1965). Assumption and fact in anthropological genetics. *Journal of the Royal Institute*, **95**, 87–103.
Sokal, R. R., Oden, N. L., Legendre, P. & Fortin, M. J. (1989). Genetic differences among language families in Europe. *American Journal of Physical Anthropology* **79**, 489–502.
Wright, S. (1946). Isolation by distance under diverse systems of mating. *Genetics*, **31**, 38–59.

19 *Population structure in the eastern Adriatic: the influence of historical processes, migration patterns, isolation and ecological pressures, and their interaction*

PAVAO RUDAN, ANITA SUJOLDŽIĆ, DIANA ŠIMIĆ, LINDA A. BENNETT AND DEREK F. ROBERTS

Introduction

In the last few decades population structure of human groups has emerged as posing some of the most interesting and provocative problems in contemporary anthropological and genetic sciences. Migration is a principal feature acting directly (though with varying intensity) on both genetic equilibrium and demography of populations, so that migration analysis is essential for the understanding of population structure at all levels (Roberts, 1988). Few human populations today remain isolated; those that are not, experience different migratory pressures. At the global level, international migration today is a major topic of economic and social concern (Appleyard, 1988), while the biological effects of human migration are of considerable importance to a wide variety of disciplines (Mascie-Taylor & Lasker, 1989) including anthropology, demography, epidemiology and genetics. A few years ago, discussing the importance of genetic structure in human microevolution, Roberts (1987) noted 'Every human population can be regarded as a continuing entity occupying a particular space. ... A population can be characterised statistically, and distinguished from other populations, by the use of parameters, its group attributes (e.g. birth rates and death rates, means and variance of metric characters, territorial density, gene frequency) which are meaningless relative to any individual. The population is permanent in relation to the individuals composing it; for the individual is born into the population, which exists before his arrival and continues to exist after his death.' This chapter presents an analysis of the results from studies of the population

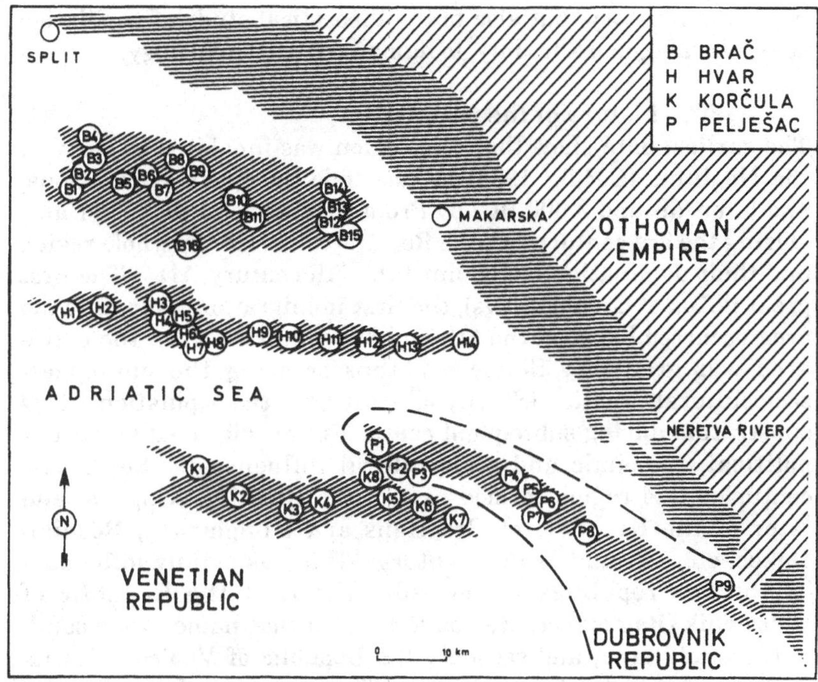

Figure 19.1. Map of the region showing locations of villages.

structure of contemporary European rural communities in the Eastern Adriatic. These rural populations are suitable for theoretical analyses of microevolution (Rudan, 1980), because of their specific ethnohistory, the known migrations that have occurred, their continuing mutual isolation, as well as the documented effects of various political, social and economic events on their biological and sociocultural formation. Research on the Eastern Adriatic coastal area of Middle Dalmatia (Republic of Croatia, Yugoslavia) was initiated in 1971 (Rudan, 1972) by the team of investigators from the Department of Anthropology (Institute for Medical Research and Occupational Health), University of Zagreb. In this area each island and peninsula, usually no more than 100 kilometers across, houses a number of small populations. Distances between some villages are no more than 3 kilometers, yet despite their close proximity they are so different in a number of distinctive population features - differences maintained, primarily, by their mutual isolation. This research on the populations of the islands of Brač, Hvar and Korcula, and the Peljesac peninsula (Figure 19.1) makes therefore, a contribution not only to the knowledge of their quite specific population history, but also to the theoretical study of population structure and, more broadly, to human population biology.

The history of the population

The earliest data show that this region was inhabited as early as the Neolithic (Čečuk, 1986) by non-Indo-European populations, and later (around 2,000 BC) by Proto-Illyrians and then Illyrians. During the 3rd century BC, the Romans colonised the whole region and their domination lasted until the 7th century AD. The first great influx of Croats (Slavs), the first immigration wave, into the area occurred between the 6th and 8th centuries when the entire area was gradually Slavicised, thus creating the biological-sociocultural Croatian (Slavic) substratum of the population. This substratum during subsequent centuries was subjected to various political, economic and sociocultural influences. For a few centuries this region formed part of the Croatian Kingdom and later of the Kingdom of Croatians and Hungarians, Bosnian Dukes, etc. From the 12th century AD it was mainly influenced by the two republics in the Adriatic, first the Republic of Dubrovnik (Ragusa) centred on the city of that name to the south of the study area, and secondly the Republic of Venice, with its centre in the north of the Adriatic. The peninsula of Pelješac was ruled by the Dubrovnik Republic from 1333; the majority of the Adriatic islands including Brač, Hvar and Korčula, as well as the mainland coastal belt, were under the rule of the Venetian Republic from 1420 and remained so until the fall of Venice in 1797 when its direct influence on the populations of the Eastern Adriatic ceased, with the appearance of Napoleon Bonaparte in this region. Control of the Pelješac peninsula by the Republic of Dubrovnik ended in 1808 when the Republic was also incorporated into Napoleon Bonaparte's 'Illyric Provinces'. This control by two contrasting political powers, over a period of approximately 400 years, was to leave its mark on the later biology, and the contrasting effects are particularly noticeable on account of the small size of the region. The peninsula of Pelješac that was dominated by Dubrovnik (Figure 19.1) lies only a mile from the island of Korčula and five miles from the island of Hvar which belonged to Venice.

During this period intensive migrations occurred in the interior of the Balkan Peninsula. The second immigration wave occurred during the Turkish wars, when many people migrated from the mainland interior of the Balkan peninsula to the Adriatic islands and nearby coastal area. These migrations continued from the 15th to 18th century, but the greatest influx of inhabitants to the islands of Brač, Hvar and Korčula, from the Makarska town area as well as the Pelješac peninsula, occurred in the mid and late 17th century during the Candian war between the Turks and the Venetians. This immigration formed a superstratum of newly arrived inhabitants, and left evidence still apparent today in clear and distinctive differences in dialect and linguistic expressions (e.g. Rubic, 1952; Rudan, 1972, 1980; Rudan *et al.*, 1987; Sujoldžić *et al.*, 1983, 1987a,b; Bennett *et al.*, 1983, 1989; Sujoldžić, 1990).

The two political forces in this period in the Adriatic had different attitudes towards newcomers. On the one hand the Venetian Republic gave them land and awarded them special privileges ('the Paštrović Privileges') according to which they were excused from serving on the Venetian galleys in times of war and were exempt from many taxes and public works; these benefits resulted in animosity, creating many sociocultural barriers, between the immigrant and indigenous populations. By marriage with natives the newcomers would lose their privileges, so that throughout the whole period of Venetian rule on the islands of Brač, Hvar and Korčula (from 1420 to 1797) there was a strong barrier to gene flow between the old population substratum and the immigrant superstratum. On the other hand, the Dubrovnik Republic (from 1333 to 1808) also permitted immigrants to settle on the Pelješac peninsula, but the socioeconomic system of feudal land ownership and its rigid control of peasants' mobility prohibited migration among the communes on the peninsula. Thus, on the islands of Brač, Hvar and Korčula, migration (gene flow) between the immigrant and indigenous population was prevented by socioeconomic privileges awarded to newcomers, while on the Pelješac peninsula gene flow was prevented by the rigid feudal system. Since these socioeconomic influences were exerted over several hundred years (approximately 15 generations) they are likely to have affected the population structure in this region. These cultural barriers enhancing inter-population isolation were disrupted by Napoleon Bonaparte's organisation of the 'Illyric Provinces' at the beginning of the 19th century, when migration movements were no longer limited, and the isolation between groups diminished (Bennett *et al.*, 1983, 1989; Rudan *et al.*, 1986, 1990).

Aim of the investigation

It is well known that geographical distance between subpopulations is an important factor which influences their biological (genetic) structure; for geographical distance can restrict, and proximity facilitate, gene flow among human populations and thus have a direct influence on the genetic kinship of the subpopulations in a region (e.g. Malecot, 1948; Roberts & Hiorns, 1962; Livingstone, 1963; Howells, 1966; Morton *et al.*, 1968; Friedlaender, 1975; Friedlaender *et al.*, 1971; Roberts, 1971, 1988; Spuhler, 1972; Morton, 1977; Relethford *et al.*, 1981; Relethford, 1985). The specific historical and cultural factors that affected the population of the Middle Dalmatian coastal area are likely to have interfered with the fundamental relationship between population structure and geographic distance, with greater kinship between closer communities. The object of the study was to establish whether such interference had occurred, and if so to examine its extent. The investigation, therefore, attempted to estimate the interaction of known historical processes, migratory patterns, isolation peculiarities, and selective ecological pressures in the formation of population structure of a part of the Eastern Adriatic.

Methods

In the holistic anthropological research on Eastern Adriatic rural populations, carried out since 1971 (Rudan, 1972, 1980) a vast amount of data has been collected from numerous villages on the islands of Brač, Hvar and Korčula, as well as from the Peljesac peninsula. These data are demographic, linguistic, anthropometric, radiographic, dermatoglyphic, physiological and genetic, e.g. parent-offspring birthplaces for migration analyses, basic vocabulary for analyses of cognate percentages, dimensions of the body and head, dimensions of metacarpal bones, quantitative finger and palm print features, cardio-respiratory measures, and blood serogenetic polymorphisms. Detailed descriptions of sample composition, measurement batteries and statistical methodology are given in, e.g. Rudan *et al.* (1987a,b, 1990).

The present report applies Malecot's (1948) 'isolation by distance' model and its parameters a and b to characterise the population structure of the islands of Brač, Hvar and Korcula and the Pelješac peninsula (Figure 19.1); Mahalanobis' D^2 (Mahalanobis, 1936) to estimate biological distances for anthropometric, physiological, morphometric and quantitative dermatoglyphic traits; Hemming's similarity measures to estimate 'linguistic distances' (Sujoldžić *et al.*, 1979). Estimates of kinship coefficients (Malecot, 1950) were based on migration data. The reduction of kinship with distance was estimated by linear regression of the logarithm of migrational kinship on geographical distance, and by non-linear regression for biological and linguistic distances. The fit of the model was tested by analysis of variance around the regression.

Results

The results of the regression analyses for the 4 areas studied are presented in Table 19.1. Sociocultural traits (migrational kinship and linguistic distances) generally fit the isolation by distance model in that the regressions are almost all significant and the majority highly so. By contrast the fit of distances assessed from traits varies in the different populations. On the islands of Brač and Korčula, physiological distances in females fit the model, although the proportion of variance accounted for on the island of Brač is very low (10%). Anthropometric body data fit the model on all except Brac and especially well on Hvar and Korčula and head dimensions on the former. Other distance estimates fit the model only on the island of Hvar. This can be interpreted as the consequence of the differentiation of the western and eastern parts of Hvar island (e.g. Rudan, 1972, 1975a,b; Rudan & Schmutzer, 1976; Rudan *et al.*, 1986; 1987a,b,c, 1990; Sujoldžić *et al.*, 1983, 1987a,b, 1989; Smolej *et al.*, 1987; Smolej-Narančić *et al.*, 1990; Šimić & Rudan, 1990; Zegura *et al.*, 1990); such differentiation, though it exists, is less pronounced in all the other investigated islands and the Pelješac peninsula.

Table 19.2 presents the estimates of the b parameter from the significant regressions in the four regions investigated. The lowest values are found on the island of Korčula, and the highest on the island of Hvar and the Pelješac peninsula. But within each area there are differences in the values estimated from different biological and sociocultural traits.

Table 19.3 presents the statistically significant correlations of biological and linguistic distances with migrational kinship (K) and geographical distances (G) (Mantel's test). Only village populations on the island of Hvar show statistically significant correlations (consistently negative) between migrational kinship and anthropometric variables of the head and body in both sexes, as well as physiological (cardiorespiratory) traits in males and metacarpal dimensions in females. Linguistic distances in the whole region, except on the island of Korčula, are negatively correlated with migrational kinship and positively with the geographical distances between individual villages, which clearly shows the influence of both geographical distance and migrational kinship on the formation of basic vocabulary. Although some biological traits and geographical distances between individual villages are significantly associated in all the areas, the majority

Table 19.1. *Values of Malecot's parameter b and the significance of the regressions*

	Brač		*Hvar*		*Korčula*		*Pelješac*	
	b	R^2	*b*	R^2	*b*	R^2	*b*	R^2
Migration - fathers	0.148	0.29***	-	-	0.060	0.51***	0.108	0.20
Migration - mothers	0.081	0.80***	-	-	0.053	0.51***	0.175	0.73***
Migration - both parents	0.096	0.75***	-	-	0.056	0.46***	0.187	0.78***
Linguistics (basic vocabulary)	0.079	0.08*	0.281	0.37**	0.097	0.40***	0.106	0.55***
Body dimensions - males (M)	0.220	0.02	0.062	0.80***	0.048	0.044***	0.330	0.30*
Body dimensions - females (F)	1.000	0.00	0.081	0.65***	0.039	0.35***	0.143	0.30*
Head dimensions - (M)	0.616	0.01	0.099	0.63***	0.049	0.18*	1.672	0.00
Head dimensions - (F)	0.491	0.00	0.101	0.70***	0.054	0.21*	1.188	0.00
Morphometry of metacarpals - (M)	0.469	0.01	0.002	0.17	1.409	0.00	0.636	0.02
Morphometry of metacarpals - (F)	1.586	0.00	0.118	0.42**	0.121	0.00	0.446	0.05
Dermatoglyphics (quantitative) - (M)	1.525	0.00	0.107	0.53***	0.314	0.01	0.356	0.03
Dermatoglyphics (quantitative) - (F)	0.262	0.04	0.265	0.28*	0.157	0.07	3.000	0.00
Dermatoglyphics (qualitative) - (M)	2.788	0.00	-	-	0.455	0.01	0.602	0.04
Dermatoglyphics (qualitative) - (F)	9.999	0.00	-	-	0.384	0.01	0.308	0.01
Physiology (cardio-respiratory) - (M)	0.311	0.04	0.521	0.01	0.093	0.08	0.183	0.20
Physiology (cardio-respiratory) - (F)	0.061	0.10**	2.358	0.09	0.066	0.33**	1.095	0.02

* $p < 0.025$
** $p < 0.010$
*** $p < 0.001$

Table 19.2. *Distribution of characters by values (significant) of Malecot's parameter b*

b	*Brač*	*Hvar*	*Korčula*	*Pelješac*
0.039			Body dimensions-f	
0.048			Body dimensions-m	
0.049			Head dimensions-m	
0.053			Kinship-m	
0.054			Head dimension-f	
0.056			Kinship-m+f	
0.060			Kinship-f	
0.061	Physiology-f			
0.062		Body dimensions-m		
0.066			Physiology-f	
0.079	Linguistics-m+f			
0.081	Kinship-m			
0.081		Body dimensions-f		
0.096	Kinship-m+f			
0.097			Linguistics-m+f	
0.099		Head dimensions-m		
0.101		Head dimensions-f		
0.106				Linguistics-m+f
0.107		Quant. dermat.-m		
0.118		Bone-f		
0.143				Body dimensions-f
0.148	Kinship-f			
0.175				Kinship-m
0.187				Kinship-m+f
0.265		Quant.dermat.-f		
0.281		Linguistics-m+f		
0.331				Body dimensions-m

of the variables analysed do not show consistently significant associations. Of the correlations with geographical distance only those for body dimensions are significant for the population of the peninsula of Pelješac. All correlations are weakest on the island of Brač, which is the closest to the mainland and has the lowest level of endogamy (37.49%), as a direct reflection of migration in the last century (Sujoldžić, 1988; Špoljar-Vržina *et al.*, 1989). The levels of endogamy on the other two islands (Hvar and Korčula where there are most correlations attaining significance) are both quite high (75.36% and 75.02%). Significant correlations are fewest on the Pelješac peninsula where endogamy is at an intermediate level.

Discussion

This research of the population structure in the Eastern Adriatic area shows well the value of the holistic approach to anthropology, advocated, for example by Harrison and Boyce (1972) on account of

Table 19.3. ***Correlations of biological and linguistic distances with migrational kinship (K) and geographical distances (G)***

	Brač		*Hvar*		*Korčula*		*Pelješac*	
	K	*G*	*K*	*G*	*K*	*G*	*K*	*G*
Linguistics (basic vocabulary)	-0.54	0.51	-0.23	0.58	-	0.68	-0.58	0.81
Body dimensions - males (M)	-	-	-0.63	0.71	-	0.63	-	0.37
Body dimensions - females (F)	-	0.28	-0.44	0.59	-	0.60	-	0.47
Head dimensions - (M)	-	0.23	-0.39	0.51	-	0.39	-	-
Head dimensions - (F)	-	0.19	-0.40	0.57	-	0.38	-	-
Morphometry of metacarpals - (M)	-	-	-	0.35	-	-	-	-
Morphometry of metacarpals - (F)	-	-	-0.39	0.50	-	-	-	-
Dermatoglyphics (quantitative) - (M)	-	-	-	0.52	-	-	-	-
Dermatoglyphics (quantitative) - (F)	-	-	-	-	-	-	-	-
Physiology (cardio-respiratory) - (M)	-	-	-0.30	-	-	-	-	-
Physiology (cardio-respiratory) - (F)	-0.21	0.42	-	-	-	0.52	-	-

Note: Statistical significance levels obtained using quadratic assignment technique

the wealth of information which it gives that helps understanding of the processes and consequences of human (micro) evolution. The investigations on almost all the rural populations in the region used the same methods on sufficiently large and representative samples. The results are intriguing, for they are not identical, and it is difficult to draw general conclusions. How to explain the results, the similarities and the differences?

The sizes of the populations under investigation from the 16th century up to now, in relation to historical economic crises and selective pressures, the pattern of migration and population isolation, can be examined thanks to the existence of the parish registers in this region (for the majority of settlements). These

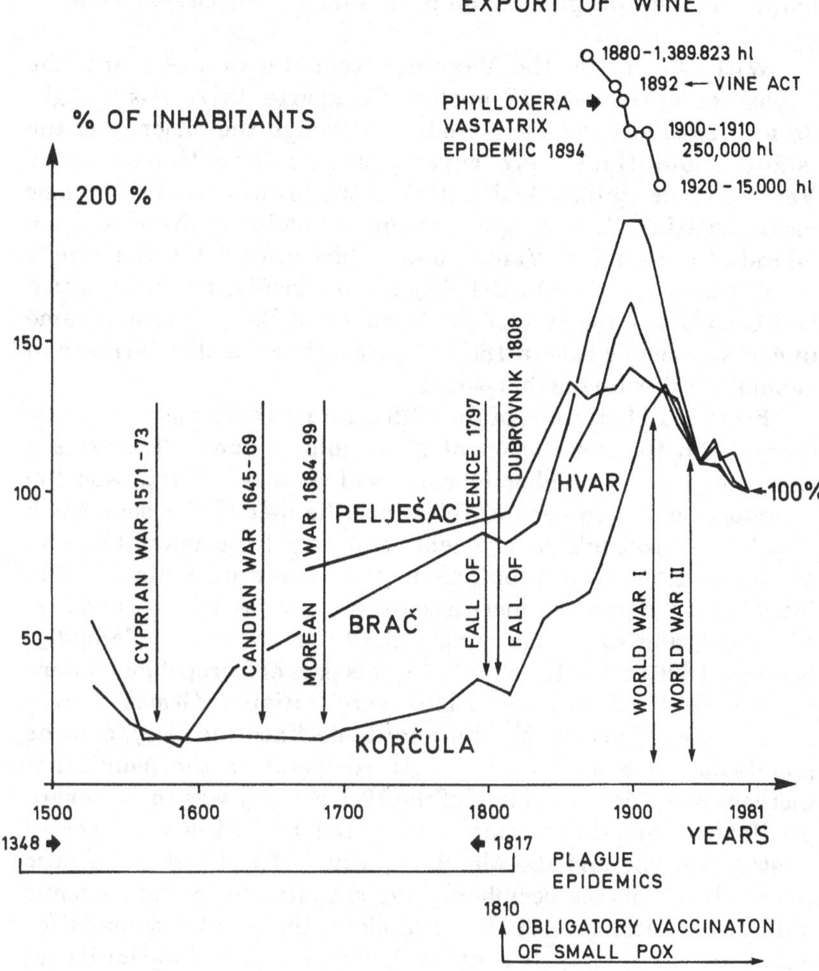

Figure 19.2. Population size and some factors influencing it.

date from the late 15th century, and it is possible to follow the size of the populations on the islands and the peninsula from 1496 (Figure 19.2). There was a decline in population during the 16th century, probably as a result of plague epidemics that swept through the area. Wars against the Turks (the Cyprian War, Candian War and Morean War) led to immigration from the mainland to the islands and the peninsula, clearly seen in the increase in population size. This increase was later and least expressed on the island of Korcula, which is the furthest from the mainland, and was greatest on Brać, which is the closest island to the coast.

With the fall of the Venetian Republic in 1797, and the imposition of the rule of Napoleon Bonaparte, there was a slight drop in population of the islands. Although the majority of the island populations were rural peasants (Croatians), under Venetian rule Italians had settled in the towns. Under the new local Croatian (Slavic) government after 1797, these left the islands to return to Venice and so accounted for the slight numerical decline. On the Pelješac peninsula, however, which had been under the rule of the Republic of Dubrovnik and came under Napoleonic rule in 1808, there continued a slow increase in population size during this period.

From 1815 to the end of the 19th century, there was a dramatic increase in the population of the whole region. This was a consequence of a number of ecological factors. First was the introduction of new food articles into the diet of the population (1816). Napoleon's government encouraged the inhabitants to plant potatoes and adopt them into their own diet. The introduction of potatoes increased the dietary possibilities and food resources, and so the carrying capacity of the area. Secondly, between 1806 and 1813 the first (in this part of Europe) mandatory annual vaccinations for smallpox were initiated (Grmek, 1964). As a result, infant and child mortality declined radically, and the population size increased. A third factor in the population increase over the second half of the 19th century was the vineyard phylloxera epidemic which devastated the wine industry of western Europe and the Mediterranean. The Eastern Adriatic region, being on the periphery, was not affected by the epidemic until much later (in 1894). Therefore, the greater demand for wines in the mid-19th century led to extensive agricultural activity in the planting of vines and increasing wine production which reached as much as 1,389,823 hectolitres in 1876. Wine was exported to Italy and the countries of western and central Europe, increasing the economic prosperity of the population. Fourth, however, the 'Wine Act' was signed in 1892 by the Austrian government (who ruled this region from the end of Napoleonic rule until 1918) and the government of Italy, by which duty was charged on the importation of wine from the Eastern Adriatic. This led to a drop in wine exports from the region, followed almost immediately by the phylloxera epidemic (1894). The consequence of these two political and economic events, the Wine Act and the phylloxera epidemic, was a drastic decrease in wine production over a period of 40 years; from 1876 to 1920 production fell from 1,389,823 hectolitres to 15,000 hectolitres a year. The resultant economic crisis led to massive emigration from this region to overseas countries. Losses in the First and Second World Wars, and increased mortality due to an epidemic of Spanish Influenza in the 1920s brought the final stage of population decline on the islands and peninsula.

The results of the present research into the population structure of this region, made on a sample born in the period from 1900 to 1970 and migration data for the same decades, lead to the following conclusions. Brač, the nearest island to the mainland and showing the lowest level of endogamy among the island groups studied (37.49%), saw an increase in immigration following increased work force needs in the last half of the 19th century (Sujoldžić, 1988; Špoljar-Vržina *et al.*, 1989), so that internal migration on that island is less than immigration from off the island. For the islands of Hvar and Korčula, similar to each other in their higher levels of endogamy (75.36% and 75.02%) and in their linear shape and orientation, immigration would predominantly have occurred from their ends nearest to the mainland and percolated from these. On the island of Korčula, the furthest from the mainland, the demographic events (reflected in population size) which affected the whole region happened one generation later. That is, in population increases and decreases, Korčula follows the same pattern as the other islands and peninsula, but 20 years later. The distinction between Korčula and the other areas is confirmed by principal component analyses of the population trends from 1857 to 1981; this yields two components which together account for 78.2% of the total variance. The first component applies to the islands of Brač, Hvar and the Pelješac peninsula, with increase in the number of inhabitants until the beginning of this century and its subsequent steady decrease until 1981 simultaneously in these 3 areas. The second component, that represents the population size on the island of Korcula, is concordant in shape but the curve shows a time lag of one generation. On the Peljesac peninsula, almost 95% of the population died in a plague epidemic in 1543, so that it was possible for new inhabitants to immigrate from the interior of the Balkan peninsula (Rudan *et al.*, 1987c; Bennett *et al.*, 1989). Once there however, no internal migration within the peninsula was permitted so that possible gene flow on the peninsula of Peljesac was very different from that in the area under Venetian rule.

Conclusion

There are striking differences in the values of parameter 'b' from

Malecot's (1948) 'isolation by distance model' between the populations studied. These can only be explained by the simultaneous analysis of both biological and ethno-historical data. All these populations live in similar environments (in terms of climate, nutrition, physical activities), in a relatively small geographical region, and they all stem initially from the same Croatian ancestral population which settled the area in the 6th to 8th centuries AD. Differences observed today therefore are not due to differences in origin or physical environment. They are to be attributed to later immigrations from various parts of the mainland, differences in within-island or peninsula migrations, differing degrees of isolation, different population sizes and experiences of economic prosperity and crises too, as well as the epidemics which devastated sometimes almost the entire population in an area. The application of the model of isolation by distance enables effects of migration intensity and geographic distance on population structure to be assessed, though with caution because of the consequences of, for example, different time-depth migrations, and of other factors which are not directly evident. Therefore, the results in this study should be considered as a first step in a broader analysis incorporating the many other elements that may contribute to the formation of contemporary human groups through isolation by distance.

Acknowledgements

The research was supported by the Smithsonian Institution, Washington, DC, USA through funds made available to the United States-Yugoslav Joint Board on Scientific and Technological Cooperation (Project SMI 862) and the Ministry of the Republic of Croatia for Science, Technology and Informatics, under the project title: 'Anthropological Research on the Population Structure of Croatia'.

References

Appleyard, R. T. (1988) *International Migration Today*. UNESCO.

Bennett, L. A., Angel, J. L., Roberts, D. F. & Rudan, P. (1983). Joint study of biological and cultural variation in Dalmatian village populations. *Collegium Antropologicum*, **7**, 195–198.

Bennett, L. A. & Rudan, P. (1989). Estimation of population structure through temporal migration analyses – example from the Island of Brac. *Collegium Antropologicum*, **13**, 85–95.

Bennett, L. A., Sujoldžić, A. & Rudan, P. (1989). Contrasts in Demographic structure and linguistic variation on the island of Korčula and the Pelješac peninsula, Yugoslavia. *Ethnologia Eurapaea*, **19**, 141–168.

Ćećuk, B. (1986) Istraživanje u spilji Kopaćini na otoku Braću i Veloj spilji na otoku Korćuli. *Obavijesti HAD*, **3**, 32–34.

Friedlaender, J. S. (1975). *Patterns of human variation*. Cambridge: Harvard University Press.
Friedlaender. J. S., Sgaramella-zonta, A., Kidd, K. K., Lai, L. V. C., Clark, P. & Walsh, R. J. (1971). Biological divergences in South-Central Bougainville: An analysis of blood polymorphism gene frequencies and anthropometric measurements using three models, and a comparison of these variables with linguistic, geographic, and migrational 'distances'. *American Journal of Human Genetics*, **23**, 253–270.
Grmek, M. D. (1964). Les conditions sanitaires et la medicine en Dalmaties sous Napoleon (1806–1813). *Extr. Biol. Med.*, 'Hors-Serie', Paris.
Harrison, G. A. & Boyce, A. J. (1972). Migration, exchange and the genetic structure of populations. In *The Structure of Human Populations*, ed. G. A. Harrison & A. J. Boyce, pp. 128–145, Oxford: Clarendon Press.
Howells, W. W. (1966). Population distances: biological linguistic, geographical and environmental. *Current Anthropology*, **7**, 531–540.
Livingstone, F. B. (1963). Blood groups and ancestry: a test case from the New Guinea Highlands. *Current Anthropology*, **4**, 541–542.
Mahalanobis, P. C. (1936). On the generalized distance in statistics. *Proceedings of the National Institute of Science, India*, **2**, 49–55.
Malecot, G. (1948). *Les mathematiques de l'heredite*. Paris: Masson.
Malecot, G. (1950). Quelques schemas probabilites sur la variabilite des populations naturelles. *Annals Université Lyon Science A* **13**, 37–60.
Mascie-Taylor, C. G. N. & Lasker G. W. (1989). *Biological aspects of human migration*. Cambridge: Cambridge University Press.
Morton, N. E. (1977). Isolation by distance in human populations. *Annals of Human Genetics*, **40**, 361–365.
Morton, N. E., Miki, C. & Yee, S. (1968). Bioassay of population structure under isolation by distance. *American Journal of Human Genetics*, **20**, 411–419.
Relethford, J. H. (1985). Isolation by distance, linguistic similarity, and the genetic structure on Bougainville Island. *American Journal of Physical Anthropology*, **66**, 317–326.
Relethford, J. H., Lee, F. C. & Crawford, M. H. (1981). Population structure and anthropometric variation in rural western Ireland: Isolation by distance and analyses of residuals. *American Journal of Physical Anthropology*, **55**, 235–245.
Roberts, D. F. (1971). The demography of Tristan da Cunha. *Population Studies*, **25**, 465–479.
Roberts, D. F. (1987). Genetic structure and differentiation of human populations. *Anthropologischer, Anzeiger*, **45**, 227–238.
Roberts, D. F. (1988). Migration and genetic change *Human, Biology*, **60**, 521–539.
Roberts, D. F. & Hiorns, R. W. (1962). The dynamics of racial intermixture. *American Journal of Human Genetics*, **14**, 261–277.
Rubić, I. (1952). *Our island in the Adriatic*. (In. Croat.). Odbor proslave 10. god. morn., Split.
Rudan, P. (1972). Etude sur les dermatoglyphes digito-palmaires des habitants de l'ile de Hvar. Doct. de Spec. en Anthrop. Biol., University of Paris VII, Paris.
Rudan, P. (1975a). Sur l'identite genetique des habitants de l'ile de Hvar. *Annals of the Institute of France*, **3**, 141–157.
Rudan, P. (1975b). The analysis of quantitative dermatoglyphic traits in the rural population of the island of Hvar, Yugoslavia. *Journal of Human Evolution*, **4**, 585–591.
Rudan, P. (1980). Microevolution studies in present populations in Yugoslavia. *Collegium Antropologicum*, **4**, 35–39.
Rudan, P., Angel, J. L., Bennett, L. A., Janićijević, B., Lethbridge, M. F., Milićić, J., Smolej-Naranćić, N. Sujoldźić, A. & Śimić, D. (1987a). Historical processes and biological structure of the populations: Example from the island of Korćula. *Acta Morphologica* (Neder - Scand), **25**, 69–82.
Rudan, P., Finka, B., Janićijević, B., Jovanović, V., Kuśec, V., Milićić, J., Miśigoj-duraković, N., Roberts, D. F., Schmutzer, L. J., Smolej-Naranćić, N., Sujoldźić, A., Śimić, D., Śimunović, P. & Śpoljar-Vrźina, S. M. (1990). Anthropological

Research of the Eastern Adriatic – *Biological and Cultural Microdifferentiation Among Rural Population of the Island of Hvar*, vol. 2 (In Croat.). Zagreb: HAD.

Rudan, P., Roberts, D. F., Janićijević, B., Smolej, N., Szirovicza, L. & Kaśtelan, A., (1986). Anthropometry and the biological structure of the Hvar population. *American Journal of Physical Anthropology*, **70**, 231–240.

Rudan, P. & Schmutzer, L. J. (1976). Dermatoglyphs of the inhabitants of the island of Hvar, Yugoslavia. *Human Heredity*, **27**, 425–434.

Rudan, P., Śimić, D. & Bennett, L. A. (1987b). Isolation by distance on the island of Koreula – Correlation analysis of distance measures, *American Journal of Physical Anthropology*, **77**, 97–104.

Rudan, P., Śimić, D., Smolej-Narańčić, N., Bennett, L. A., Janićijević, B., Jovanović, V., Lethbridge, M. F., Milićić, J., Roberts, D. F., Sujoldźić, A. & Szirovicza, L., (1987c) Isolation by distance in Middle Dalmatia – Yugoslavia. *American Journal of Physical Anthropology*, **74**, 417–426.

Śimić, D. & Rudan, P. (1990). Isolation by distance and correlation analyses of distance measures in the study of population structure (example from the island of Hvar). *Human Biology*, **62**, 113–130.

Smolej, N., Angel, J. L., Bennett, L. A., Roberts, D. F. & Rudan, P. (1987). Physiological variation and population structure of the island of Korćula, Yugoslavia. *Human Biology*, **59**, 667–685.

Smolej-Narańčić, N., Rudan, P. & Bennett, L. A. (1990). Anthropometry and the biological structure of the population: example from the Island of Brać. In *Approche pluri-disciplinaire des isolats humains*, pp. 243–270. ed. A. Chaventre D. F., Roberts, Paris: INED.

Spuhler, J. N. (1976). Genetic, linguistic and geographical distances in Native North America. In *The assessment of population affinities in Man*, ed. J. S., Weiner & J. Huizinga, pp. 72–95. Oxford: Clarendon Press.

Sujoldźić, A. (1988). The population structure of the island of Brać: A dermatoglyphic and migrational analysis. *Collegium Antropologicum*, **12**, 329–351.

Sujoldźić, A. (1990). The analysis of population history and cultural (linguistic) microevolution of the Slavic settlements in Molise, Italy. *Homo*, **41**, 1–15.

Sujoldźić, A., Finka, B., Śimmunović, P. & Rudan, P. (1987a). Language and origin of the inhabitants of Slavic settlements in the region of Molise, Italy. (In Croat.) Rasprave zavoda za jezik, **13**, 117–145.

Sujoldźić, A., Jovanović, V., Angel, J. L., Bennett, L. A., Roberts, D. F. & Rudan, P., (1989). Migration within the island of Korćula, Yugoslavia, *Annals of Human Biology*, **16**, 483–493.

Sujoldźić, A., Szirovicza, L., Momirović, K., Finka, B., Moguś, M., Śimunović, P. & Rudan, P. (1979). Application of taxonomic algorithm on non-numeric variables in the study of linguistic microevolution. (In Croat.). *Rasprave Zavoda za jezik*, **4–5**, 61–68.

Sujoldźić, A., Szirovicza, L., Śimunović, P., Finka, B., Roberts, D. F. & Rudan, P., (1983). Linguistic distances on the island of Hvar. (In Croat.). *Rasprave Zavoda za jezik*, **8–9**, 197–214.

Sujoldźić, A., Śimunović, P., Finka, B., Bennett, L. A., Roberts, D. F., Angel, J. L. & Rudan, P. (1987b). Linguistic microdifferentiation on the island of Korćula. *Anthropological Linguistics*, **28**, 405–432.

Bennett, L. A. & Rudan, P. (1990). Population Structure of the Pelješac Peninsula, Yugoslavia. *Human Biology*, **62**, 173–194.

20 *Diabetes and diabetic macroangiopathy in Japanese-Americans*

HITOSHI HARA, GENSHI EGUSA, KIMINORI YAMANE, MICHIO YAMAKIDO

It is generally thought that in Japanese the incidence of cerebrovascular diseases was formerly high and the incidence of ischemic heart diseases and diabetes low (Kays *et al.*, 1966; WHO, 1980), but that recent westernisation of the Japanese life style, particularly diet, has brought changes. To clarify the influences of westernisation of life style on diseases in Japanese, in a recent survey Japanese-Americans, in whom westernisation of life style occurred earlier and more intensively, were compared with Japanese living in Japan, where these changes have been later and less pronounced.

The survey of Japanese-Americans was started in 1970 in Hilo City and Kona District, Hawaii. In 1978, the survey in Los Angeles was started. The results of the first of these surveys of Japanese-Americans, and of Japanese in Hiroshima Prefecture, have already been reported (Kawate *et al.*, 1979). This paper reports the results of subsequent surveys in Hawaii, Los Angeles and Hiroshima, placing emphasis on the characteristics of diabetes mellitus among Japanese-Americans.

Background of Japanese migrants and their offspring

Emigration of Japanese to Hawaii began around 1885 in reply to the request from the Hawaiian Dynasty of those days to alleviate the shortage of labour for the sugar cane plantations. After World War II, the social status of these migrants improved sharply. At present, second- and third-generation Japanese-Americans are active in various occupations in Hawaii. In Hilo City on the island of Hawaii where the survey was conducted, 14,000 Japanese-Americans live, exceeding the population of white Americans in that city. The Japanese migrants living in Hilo City came from various prefectures of Japan, and have organised societies of people from each prefecture. Of these prefecture-oriented societies, the society for emigrants from Hiroshima Prefecture, which was founded more than a century ago, is the largest with 1,279 members in 1986. Of these, 1,026 (80.2%) were surveyed. The subjects of this study in Hawaii thus were members of the Society of Emigrants from Hiroshima Prefecture and their families.

Emigration of Japanese to Los Angeles began later than emigration to Hawaii. Unlike Japanese-Americans in Hawaii, those in Los Angeles account for only a small percentage of all citizens in this multi-racial city. A relatively high percentage of Japanese-Americans examined in Los Angeles were engaged in agriculture (farming, horticulture or nurserymen). In this area also a society of emigrants from Hiroshima (South California Society of People from Hiroshima Prefecture) was organised; it had 1,743 members in 1982, and 1,232 (70.7%) were included in this survey. Again the subjects studied in Los Angeles were members of this society and their families.

In Hiroshima Prefecture (population 2.7 million), individuals who received health checks for adult diseases (performed by us in various districts of Hiroshima Prefecture) at almost the same time as our surveys of Japanese-Americans were randomly selected for this study.

Subjects

During the period from 1978 to 1989, eight medical surveys were carried out in Hawaii, Los Angeles or Hiroshima, and the subjects of this study were those attending for the first time. There were 643 Japanese-Americans in Hawaii, 1,232 Japanese-Americans in Los Angeles and 2,275 Japanese in Hiroshima. The subjects in Hiroshima contained more females and slightly fewer elderly individuals than in Hawaii and Los Angeles. The percentage of first-generation immigrants among all subjects was higher in Los Angeles (36%) than in Hawaii (19%).

Methods

The method of this medical survey has been reported elsewhere (Kawate *et al.*, 1979). The same physicians and nutritionists performed the examinations in the three areas.

The nutritional survey was carried out by individual interviews, in which questions were asked on the amount and frequency of intake of various foods over a relatively long period, and a microcomputer was used for calculation of mean nutrient intakes. Reported physical activity levels were rated on a 3-point scale (based on the Labor Intensity Table, Ministry of Health and Welfare [Matsushima, 1970]) after interviews of past and present occupation. Because of recent trends of mechanisation in workshops and the high percentage of elderly retired individuals in Japanese-Americans, a final judgement of physical activity level was made, taking into consideration the degree of daily physical labour and the duration of standing work per day. From the measurements of height and weight of each subject, the Body Mass Index (BMI, kg/m^2) was calculated.

In all subjects, venous blood was sampled after an overnight fast, and oral glucose tolerance tests were performed. The specimens were stored in dry ice after separation of serum and then transported to Hiroshima by air. All specimens from the three areas were simultaneously assayed at the same laboratory in Hiroshima. LDL-cholesterol was calculated using the equation of Friedewald (Friedewald *et al.*, 1972). Causes of death based on the reports from the Hawaii State Public Health Bureau were analysed.

Results

In each sex, all three areas showed similar mean daily energy consumption (Table 20.1). A more detailed analysis, dividing proteins and fats into those of animal and vegetable origin and carbohydrates into simple (fructose and sugar) and complex, showed that Japanese-Americans took more protein of animal than of vegetable origin. Their consumption of fat of animal origin was approximately twice that of Japanese in Hiroshima. In Japanese-Americans consumption of simple carbohydrate was 1.5-2.0 times, but that of complex carbohydrate only one half to two thirds, that in Hiroshima citizens. Evidently, Japanese-Americans had a westernised diet style (rich in animal fat and simple carbohydrate and poor in complex carbohydrate) as compared to Japanese in Hiroshima.

There was no difference among the three areas in the percentage of males or females with 'light' physical activity, but there was a lower percentage of individuals with 'heavy' physical activity in Japanese-Americans (Kawate *et al.*, 1979; Hara *et al.*, 1983).

When the degree of obesity was compared using the Body Mass Index, the percentage of obese males (BMI over 25) was significantly higher in Japanese-American than in Hiroshima citizens, but in females there was no significant difference among the three areas.

Table 20.1. *Mean nutrient intake*

Nutrients	*Male*			*Female*		
	Hawaii	*LA*	*Hiroshima*	*Hawaii*	*LA*	*Hiroshima*
No. of cases	434	367	636	574	464	657
Total energy (Kcal)	2243	2421	2313	1701	1754	1925
Total protein (g)	82	88	75	66	68	68
Animal origin	49	54	37	40	40	35
Vegetable origin	33	34	38	26	27	33
Total fat (g)	78	86	53	61	64	52
Animal origin	40	46	26	31	32	25
Vegetable origin	37	40	27	30	32	27
P/S ratio	0.9	0.9	1.2	1.0	0.9	1.2
Total carbohydrate (g)	293	301	322	227	230	284
Simple	89	100	57	83	86	67
Complex	204	201	266	144	146	217
Dietary fibre (g)	14	13	11	12	13	12
Salt (g)	11	12	13	10	11	12

The crude prevalence of diabetes as diagnosed using WHO criteria (WHO, 1980) in individuals aged over 40 years was 21.6, 15.6 and 5.8% for Hawaii, Los Angeles and Hiroshima citizens, respectively. Corrected by the direct method based on sex, age and obesity distributions in Hiroshima subjects, the prevalence was 14.3, 10.7 and 5.8% for Hawaii, Los Angeles and Hiroshima citizens. When corrected based on sex and age distribution for the population of Hiroshima Prefecture, the prevalence was 18.3, 13.4 and 6.2% for the three areas, respectively. The prevalence of diabetes among Japanese-Americans was twice or three times that in Hiroshima citizens.

Analysed in relation to generation, the prevalence of diabetes in individuals aged over 40 years was 15.9% for first-generation Japanese-Americans and 18.9% for second- to fourth-generation Japanese-Americans. Corrected for sex, age and obesity distributions for Hiroshima subjects diabetes prevalence was 8.2% for first generation and 13.9% for second to fourth generation.

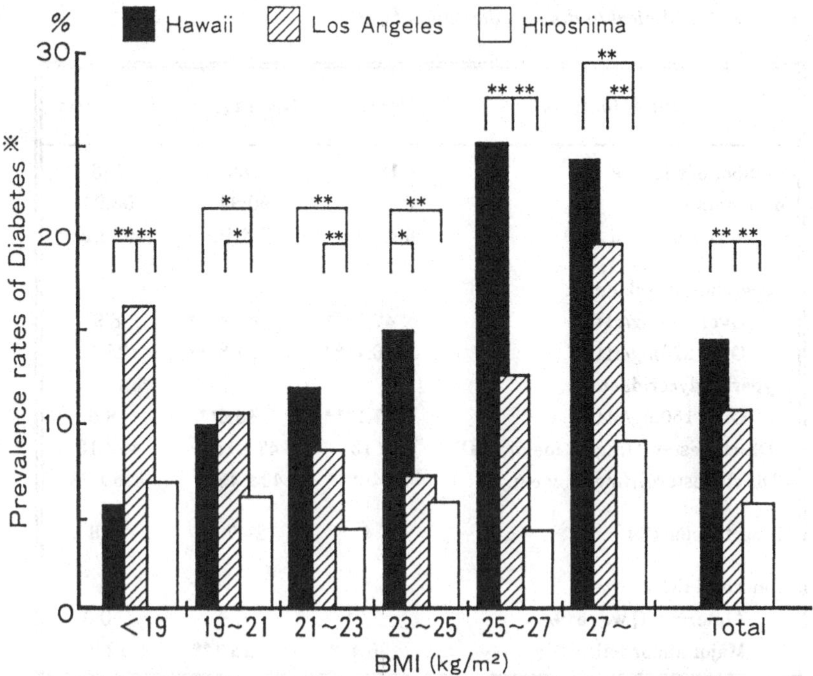

Figure 20.1. Age-sex specific prevalence of diabetes by BMI.

Thus, the prevalence of diabetes in second- to fourth-generation Japanese-Americans was significantly higher than in Hiroshima.

Figure 20.1 shows the age- and sex-corrected prevalence of diabetes for 6 groups of subjects by the obesity level in the three areas. When compared within the same obesity level, the prevalence of diabetes was significantly higher in Japanese-Americans than in Hiroshima citizens.

Table 20.2 shows a comparison of the laboratory data among diabetics in the 3 areas. The prevalence of hypercholesterolaemia, hypertriglyceridaemia and ischemic ECG abnormalities (Pooling Project Research Group, 1978) were significantly high in Japanese-Americans with diabetes, while there was no difference among areas in the prevalence of hypo-HDL-cholesterolaemia or hypertension as defined by WHO criteria (WHO, 1978).

The sex- and age-corrected prevalence of diabetes of males in Hawaii and Los Angeles divided by physical activity levels into three groups shows the prevalence of diabetes to be significantly lower in the heavy physical activity group (Table 20.3).

Table 20.2. *Clinical findings in diabetic subjects*

Clinical findings	*Hawaii*	*Los Angeles*	*Hiroshima*
Number of subjects	189	183	148
Male:Female	87:102	90:93	58:90
Age in years (mean ± SD)	66 ± 10	65 ± 10	59 ± 9
Hypercholesterolaemia:			
Over 250mg/dl (%)	47.1***	45.4***	8.9
Over 220mg/dl (%)	76.7***	67.8***	28.1
Hypertriglyceridaemia:			
Over 150mg/dl (%)	57.1***	48.6***	18.5
HDL-cholesterol (mg/dl, mean ± SD)	46 ± 13	47 ± 15	46 ± 13
LDL-cholesterol† (mg/dl, mean ± SD)	144 ± 40***	143 ± 49**	128 ± 38
Hypertension (%)	34.4	24.7	23.8
Abnormal ECG*			
Abnormal Q wave (%)	3.2**	0.6	0.0
Major abnormality (%)	25.4**	25.1**	12.8

† Calculated by the Friedewald's formula.

***$p < 0.005$; **$p < 0.01$; *$p < 0.05$ (vs Hiroshima)

* Pooling project's criteria by Minnesota code. Abnormal Q wave: I-1,2; Major abnormality: IV-1,2; V-1,2; VII-1,2,4; VIII-1,3.

Table 20.3. *Effect of levels of physical activity on prevalence of diabetes*

*Physical Activity**		*Age (years) (M ± SD)*	*BMI (Kg/m2) (M ± SD)*	*Energy intake (Kcal/day) (M ± SD)*	*Age-sex adjusted prevalence of diabetes (%)*
Light	Male	59 ± 11	24.5 ± 3.1	2238	10.1
	Female	59 ± 9	23.9 ± 32	1686	
Moderate	Male	61 ± 10	24.5 ± 3.8	2248	10.5
	Female	59 ± 9	22.7 ± 3.3	1786	
Heavy	Male	59 ± 8	23.5 ± 2.9	2852	5.2
	Female	57 ± 6	22.2 ± 2.4	2052	

*Classified by the criteria of occupational classification tabulated by the Japanese Ministry of Health and Welfare, and averaged daily standing hours from the postural history.

BMI: Body Mass Index

Table 20.4. ***Effects of dietary carbohydrate-to-fat intake ratio on clinical findings in Japanese-Americans***

Group	*A*	*B*	*C*	*D*
Carbohydrate (%)	60	55	50	45
Fat (%)	25	30	35	40
Number of cases (%)	158	204	179	117
Age (M ± SD)	60 ± 7	58 ± 8	59 ± 8	59 ± 7
BMI (M ± SD)	23 ± 1	22 ± 1	23 ± 2	23 ± 2
Diabetes (%)	7.0	6.4	13.4*	13.7*
Hypercholesterolaemia (%)	32.1	27.5	33.7	37.3
Hypertriglyceridaemia (%)	35.9	32.9	35.4	24.6
Low HDL cholesterolaemia (%)	23.1	19.4	20.2	14.5
Hypertension (%)	13.2	12.1	11.1	7.6

*$p < 0.05$ (vs Group B)
BMI: Body Mass Index (kg/m^2)
Hypercholesterolaemia: Total serum cholesterol >250mg/dl
Hypertriglyceridaemia: Serum triglyceride >150mg/dl
Low HDL cholesterolaemia: Serum HDL-cholesterol <40mg/dl
Diabetes, Hypertension: WHO criteria

The prevalence of diabetes for 4 groups of Japanese-Americans divided by the carbohydrate-to-fat ratio in energy intake (Table 20.4) was highest for Group D (highest fat intake) and lowest for Group B.

Aortic pulse wave velocity (PWV), which allows non-invasive diagnosis of arteriosclerosis (Hasegawa, 1970; Hara *et al.*, 1986), was determined in 552 Japanese-Americans in Hawaii. Comparison of the data by age group with those from 40,665 healthy Japanese and 231 ambulatory patients with diabetes visiting the Hiroshima University Hospital showed that the mean PWV for Japanese-Americans in Hawaii aged over 50 years was significantly higher than for healthy Japanese of the same age group and was close to the mean for Japanese diabetics (Figure 20.2). For 97 diabetic Japanese-Americans in Hawaii the mean ± SD of PWV was 10.9 ± 2.2m/sec which was significantly higher than the figures for 173 Hiroshima University Hospital outpatients (9.4 ± 2.1; $p < 0.001$).

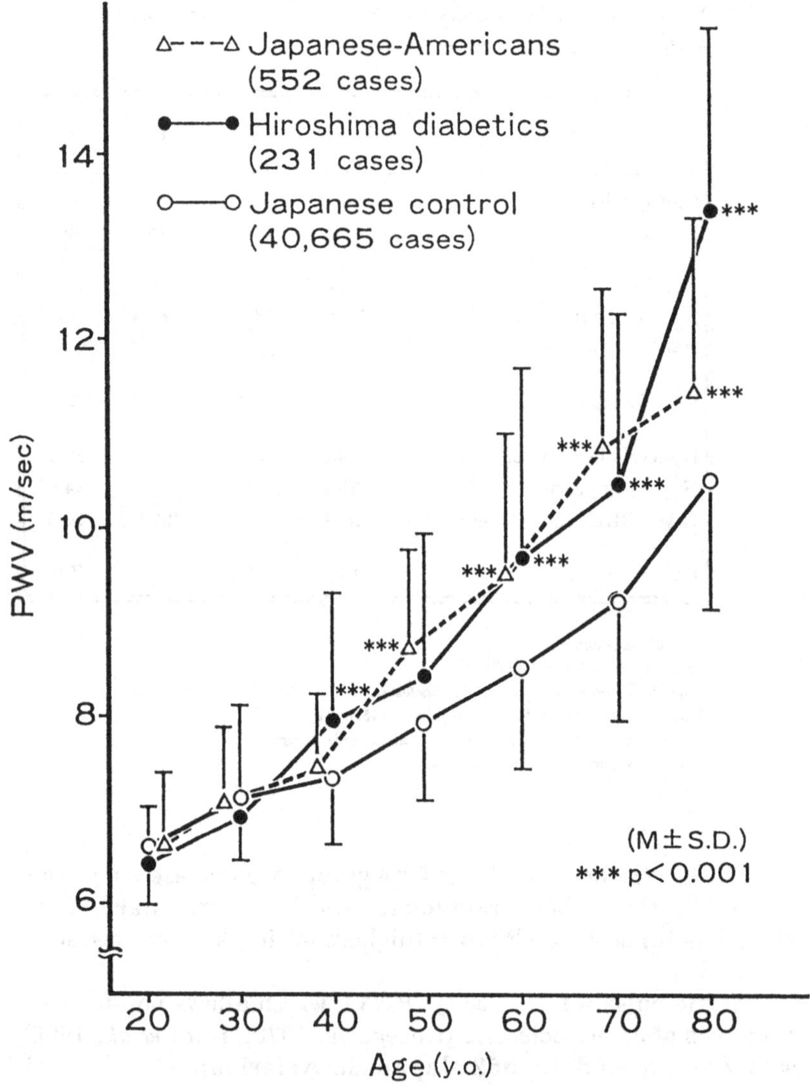

Figure 20.2. Age-related changes of PWV (aortic pulse wave velocity) in Japanese controls, Japanese-Americans in Hawaii and diabetics in Hiroshima University Hospital.

Serum insulin was determined before, one hour after and 2 hours after the glucose load. Non-diabetic, non-obese individuals (BMI below 25) were divided into 6 groups according to obesity levels, with blood glucose level in a 50g oral glucose tolerance test approximately matched for all groups. The mean serum insulin

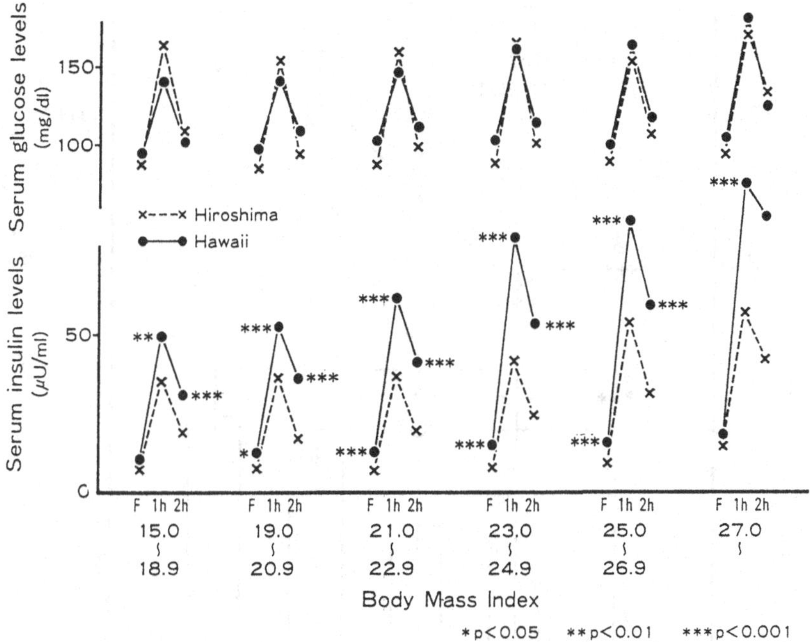

Figure 20.3. Mean serum insulin levels during 50g OGTT in non-diabetic subjects in relation to BMI.

level for each group was compared with that for Hawaii Japanese-Americans and Hiroshima citizens (Figure 20.3). At most obesity levels, Japanese-Americans showed a high prevalence of fasting hyperinsulinaemia and an elevated insulin reaction to glucose load as compared to Hiroshima citizens. Similar results were obtained when compared with Los Angeles and Hiroshima in a 75g oral glucose tolerance test.

The Japanese-Americans were divided into 5 groups according to ΣIRI (sum of serum insulin levels during the glucose tolerance test). For each group was calculated the mean PWV and ΔPWV (the difference between the PWV value for individual Japanese-Americans and the mean for healthy Japanese of the same age group). As ΣIRI rose, PWV and ΔPWV increased (Figure 20.4).

Based on death registrations for Hawaii Island and Hawaii State, provided by the Hawaii State Public Health Bureau, the causes of death in white Americans and Japanese-Americans with diabetes were analysed. Those for whom diabetes was included as one of the three possible causes of death listed on death certificates were regarded as diabetics and included in this analysis. The percentage of diabetics among total deaths of Japanese-Americans

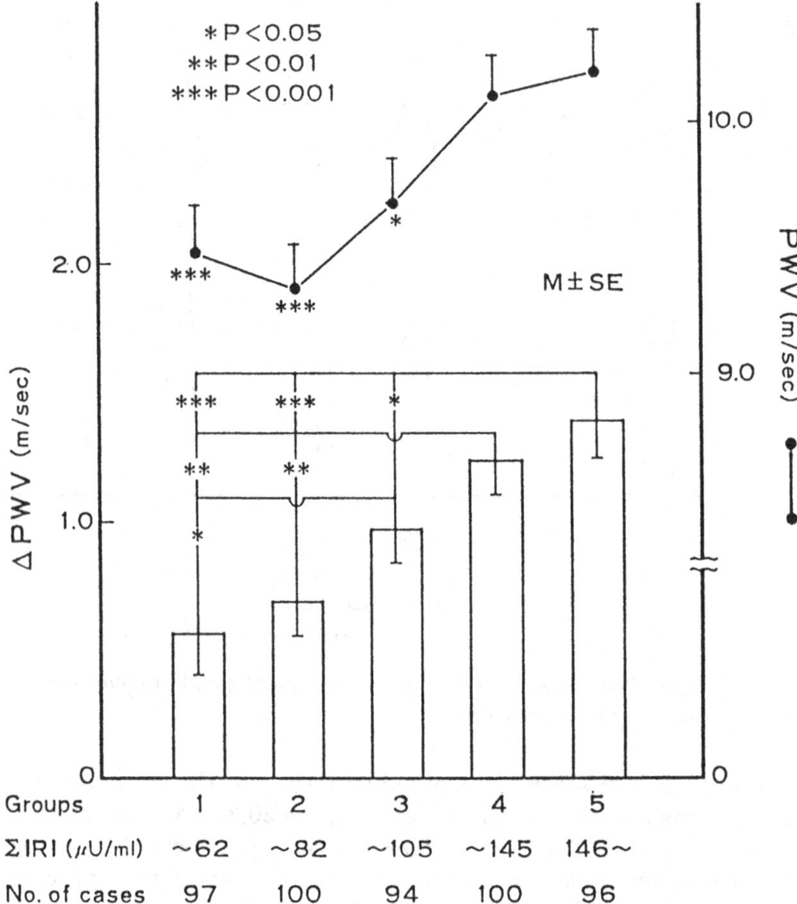

Figure 20.4. Effects of hyperinsulinaemia on PWV and ΔPWV. ΔPWV = difference between PWV values for individual Japanese-Americans and mean PWV for healthy Japanese of the same age group.

was about half that for white Hawaiians (2.9% vs 7.6%) in the 1950s, became almost equal to that for white Hawaiians in the 1960s (5.7% vs 6.1%), and exceeded it in the 1970s (10.2% vs. 6.6%). Also in the 1980s, the percentage for Japanese-Americans was higher than that for white Hawaiians (10.2% vs. 4.8%).

Comparison was made of the percentages of Japanese-American diabetics who died of vascular diseases with those in mainland Americans and Japanese. The mainland American figures derived from the Joslin Clinic data of Entmacher *et al.* (1985) and the Japanese from the nation-wide surveys by questionnaire by Sakamoto and Kosaka (1981) and of autopsied

Table 20.5. *Frequency of vascular disease in deceased diabetic subjects*

	USA	*State of Hawaii*		*Japan*	
	*Clinical diagnosis Joslin Clinic**	*Caucasian*	*Japanese-American*	*Clinical diagnosis (Sakamoto)*	*Autopsy cases (Goto)*
Years	*1969-1979*	*1970-1987*	*1970-1987*	*1971-980*	*1981-1982*
No. cases	*4290*	*1422*	*2551*	*9737*	*1939*
Total	75.6	71.4	72.8	41.5	47.0
Brain	11.1	10.9	15.9	16.4	13.7
Cardiac	54.5	54.5	50.1	15.9	
IHD		47.2	44.0	12.3	14.4
Renal	7.0	2.3	3.6	12.8	17.0

*Reported by Entmacher *et al.* (1985)

cases by Goto (1987) (Table 20.5). The percentage of deaths due to ischemic heart diseases in Japanese-Americans with diabetes was high and comparable to that for white Americans, clearly contrasting with that in Japanese.

Discussion

This comparison of results of medical surveys of Japanese-Americans in Hawaii Island and Los Angeles with those of Japanese in Hiroshima Prefecture disclosed a higher prevalence of diabetes among Japanese-Americans (about twice or three times that of Hiroshima citizens even after correction for sex, age and obesity level). The prevalence of diabetes was higher in second- to fourth-generation Japanese-Americans (who were born in America) than in first-generation Japanese-Americans (emigrants from Japan). The high prevalence of diabetes among Japanese-Americans was also demonstrated by death statistics from the Hawaii State Public Health Bureau and by the recent epidemiological study of Fujimoto *et al.* (1987) of Japanese-Americans in Seattle.

Onset of diabetes is thought to involve two major groups of factors (genetic and environmental), and its prevalence markedly varies between nations or races (WHO, 1980). In the present study, the Japanese-Americans examined were ethnically pure

Japanese who seem to differ genetically little if at all from Hiroshima citizens because many of them originated from this prefecture. Therefore, the high prevalence of diabetes among Japanese-Americans is to be attributed to environmental rather than genetic factors.

At present, obesity is regarded by many as the most important etiological factor in the onset of diabetes (West & Kalbfleisch, 1971; Zimmet, 1982). The observed percentage of obese individuals was higher in Japanese-Americans (in particular males) than in Hiroshima citizens, but the excess prevalence of diabetes in Japanese-Americans over that in Hiroshima citizens remained when individuals at the same obesity levels were compared (Figure 20.1), suggesting that diabetes in Japanese-Americans involves some factors other than obesity.

Following this finding, differences were explored in environmental factors (other than obesity) between Japanese-Americans and Hiroshima citizens. Analysis of reported diet and physical activity strongly suggested that the high prevalence of diabetes among Japanese-Americans is associated with low physical activity and ingestion of foods rich in animal fat and simple carbohydrate and poor in complex carbohydrate. But further important information in Japanese-Americans was provided by comparison of serum insulin levels in Japanese-Americans and Hiroshima citizens of the same obesity level and with matched blood glucose levels. The higher prevalence of hyperinsulinaemia and elevated insulin response to glucose load in Japanese-Americans suggest that insulin resistance of peripheral tissue was enhanced in Japanese-Americans. Perhaps the enhanced peripheral tissue insulin resistance is a major factor responsible for the high prevalence of diabetes among Japanese-Americans. Low physical activity and intake of high fat and low carbohydrate foods, are important factors causing enhancement of peripheral tissue insulin resistance (Lipman *et al.*, 1972; Taylor *et al.*, 1984; Chen *et al.*, 1988; Storlien *et al.*, 1986).

As regards the features of macroangiopathy among diabetic Japanese-Americans, their significantly higher prevalence of hyperlipidaemia and mean LDL-cholesterol level may be suggested as contributory factors in their high mortality due to ischemic heart diseases (Table 20.5). A diet rich in fats of animal origin is thought to elevate blood LDL-cholesterol level (Baudet *et al.*, 1984). The aortic pulse wave velocity findings and their good positive correlation with totaled insulin levels during the glucose tolerance test (Figure 20.4) add further information. Based on all the results from the present study, and the knowledge that hyper-

LDL-cholesterolaemia and hyperinsulinaemia are two major risk factors for arteriosclerosis (Gordon *et al.*, 1977; Goldstein, 1973; Stout, 1973; Janka, 1986), the high mortality of diabetic Japanese-Americans from ischemic heart disease seems to be chiefly associated with hyperlipidaemia and hyper-insulinaemia.

References

Baudet, M. F., Dachet, C., Lasserre, M., Esteva, O., Jacotot, B. (1984). Modification in the composition and metabolic properties of human low density and high density lipoproteins by different dietary fats. *Journal of Lipid Research*, **25**, 456–468.

Chen, M., Bergman, R. N. & Porte, D. Jr. (1988). Insulin resistance and ß-cell disfunction in aging: The importance of dietary carbohydrate. *Journal of Clinical Endocrinology & Metabolism*, **67**, 951–957.

Entmacher, P. S., Krall, L. P. & Kranczer S. N. (1985). Diabetic mortality from vital statistics. In *Joslin's Diabetes Mellitus*, 12th edn, ed. A. Marble, L. P. Krall, R. F. Bradley, A. R. Christlieb, J. S. Soeldner 278–297. Philadelphia: Lea & Febiger.

Friedewald, T., Levy, R. I. & Fredrichso, D. S. (1972). Estimation of the concentration of low-density lipoprotein cholesterol in plasma, without use of the preparative ultracentrifuge. *Clincal Chemistry*, **18**, 499–502.

Fujimoto, W. Y., Leonetti, D. L., Kinyoun, J. L., Newell-Morris, L., Shuman, W. P., Stolov, W. C. & Wahl, P. W. (1987). Prevalence of diabetes mellitus and impaired glucose tolerance among second-generation Japanese-American men. *Diabetes*, **36**, 721–729.

Goldstein, L. (1973). Hyperlipidemia in coronary heart disease. *Journal of Clincal Investigation*, **52**, 1533–1544.

Gordon, T., Castelli, W. P., Hjortland, M. C., Kannel, W. B. & Dawber, T. R. (1977). Predicting coronary heart disease in middle-aged and older persons. The Framingham Study. *Journal of the American Medical Association*, **238**, 497–499.

Goto, Y. (1987). Changes of causes of death among diabetes in Japan. *Diagnosis and Treatment*, **73**, 335.

Hara, H., Morita, N., Ogawa, J., Egusa, G., Kobuke, A., Kubo, K., Okubo, M., Tanabe, Y., Takayama, S., Matsumoto, Y., Yamakido, M. & Nishimoto, Y., (1986). Pulse wave velocity (PWV) in diabetics and Japanese-Americans in Hawaii. *Journal of the Japanese Diabetic Society*, **29**, 737–748.

Hasegawa, M. (1970). Basic study on human aortic pulse wave velocity. *Jikei Medical Journal*, **85**, 742–760.

Janka, U. (1986). Risk factors of cardiovascular complications in diabetes mellitus. In *Diabetes* 1985, ed. M. Serrano-Rios & J., Lefebvre, pp. 717–724. Elsevier: Amsterdam.

Kawate, R., Yamakido, M., Nishimoto, Y., Bennett, P. H., Hamman, R. F. & Knowler, W. C. (1979). Diabetes mellitus and its vascular complications in Japanese migrants on the island of Hawaii. *Diabetes Care*, **2**, 161–170.

Kays, A., Aravanis, C., Blackburn, H. W., Van Buchem, F. S. P., Buzina, R., Djordjevic, B. S., Dontas, A. S., Fidanza, F., Karvonen, M. J., Kimura, N., Lekos, D., Monti, M., Puddu, V. & Taylor, H. L. (1966). Epidemiological studies related to coronary heart disease: characteristics of men aged 40–59 in seven countries. *Acta Med. Scand. Supple.* **460**.

Lipman, R. L., Raskin, P., Love, T., Triebwasser, J., Lecocq, F. R. & Schnure, J. J., (1972). Glucose intolerance during decreased physical activity in man. *Diabetes*, **21**, 101–107.

Matsushima, S. (1970). In *Current status of physical strength in Japan*, ed. S., Matsushima p.171. Tokyo: Daiichi Houki Shuppan Co.

Pooling Project Research Group (1978). Relationship of blood pressure, serum cholesterol, smoking habit, relative weight and ECG abnormalities to incidence of major coronary events. Final report of the Pooling Project. *Journal of Chronic Disease*, **31**, 201–306.

Sakamoto, N. & Kosaka, K. (1981). Diabetes and macroangiopathy. *Journal of the Japanese Diabetic Society*, **24**, 1146–1147.

Storlien, L. H., James, D. E., Burleigh, K. M., Chisholim, D. J. & Kraegen, E. W. (1986). Fat feeding causes widespread in vivo insulin resistance, decreased energy expenditure, and obesity in rats. *American Journal of Physiology*, **251**, E576–E583.

Stout, W. (1973). The role of insulin in the development of atherosclerosis. In *Vascular and neurological changed in early diabetes*, ed. R. Camerini-Davulos & H. Cole, pp. 41–47. New York: Academic Press.

Taylor, R., Ram, P., Zimmet, P., Raper, L. R. & Ringrose, H. (1984). Physical activity and prevalence of diabetes in Melanesian and Indian men in Fiji. *Diabetologia*, **27**, 578–582.

West, K. & Kalbfleisch, J. M. (1971). Influence of nutritional factors on prevalence of diabetes. *Diabetes*, **20**, 99–108.

W.H.O., Expert Committee. Technical Report Series 628. (1978). *Arterial hypertension*. Geneva: W. H. O.

W.H.O. Expert Committee on Diabetes Mellitus (1980). Second Report. *W.H.O. Technical Report Series*, No. 646.

Zimmet, P. (1982). Type 2 (non-insulin-dependent) diabetes – An epidemiological overview. *Diabetologia*, **22**, 399–411.

21 *Diabetes and westernisation in Japanese migrants*

YASUNORI KANAZAWA, MAGID IUNES AND WILFRED Y. FUJIMOTO

Introduction

Type 2 diabetes, or non insulin-dependent diabetes mellitus (NIDDM), is less frequent in Japan than in the United States. It is a heritable disease as demonstrated by a very high concordance rate (95%) in Japanese and European identical twins (Kuzuya *et al.*, 1987; Barnett *et al.*, 1981). However, it is also well known that many environmental factors contribute to the pathogenesis of NIDDM. Thus, analyses of both genetic and environmental factors and their interactions may be expected to promote further understanding of the pathogenesis of diabetes and provide knowledge useful to its prevention.

This paper presents epidemiological data about type 2 diabetes in Japanese in Japan as compared to Japanese migrants to the United States and Brazil. Since the modernisation of life style that is occurring in Japan may have a strong influence upon the incidence of diabetes there, a study of Japanese migrants who have already experienced these changes elsewhere associated with westernisation throughout their life may show the rates for diabetes to be expected when Japan is fully westernised. This paper also describes the difficulties encountered in a study of relatives in Japan of Seattle Nisei (second generation Japanese-American) who had been studied by Fujimoto *et al.* (1987a,b).

Epidemiology of diabetes mellitus in Japan

There have been few population-based studies on prevalence and incidence of diabetes mellitus in Japan. Since government health insurance covers almost all of the Japanese population, statistics of the Ministry of Health and Welfare can be used to estimate the prevalence of diabetes (Health and Welfare Statistics Association, 1989). In 1986, there were 920,400 patients 45 to 75 years old who visited hospital or a doctor's office with the diagnosis of diabetes

Table 21.1. *Prevalence (%) of type 2 diabetes from two population-based studies from rural and urban areas of Japan* (Shigeta *et al.*, 1983; Kitazawa *et al.*, 1983)

	Sex	*Diabetes mellitus*	*Impaired glucose tolerance*
Rural (Shiga area)	male	3.0	-
	female	1.2	-
Urban (Tokyo area)	male	4.0	18.3
	female	2.0	20.0

Table 21.2. *Prevalence (%) of diabetes and impaired glucose tolerance in 45-75 year old Nisei (Seattle study)*

	Male	*Female*
Already diagnosed diabetes	13	8
Already and newly diagnosed diabetes	20	16
Impaired glucose tolerance	36	40

Table 21.3. *The increasing numbers of diabetic patients and the calculated annual rate of increase (%)*

Year	*Number of patients (x1000)*	*Annual rate of increase*
1955	17.9	
1960	37.4	16
1965	78.6	16
1970	166.0	16
1975	206.2	4.5
1980	503.7	19
1985	738.4	8

mellitus. Since the total size of this age group was 36,170,000, the prevalence of diabetes was 2.54%.

Two surveys were carried out, one rural and one urban. The rural survey was carried out in Aito village in Shiga Prefecture by investigators of Shiga Medical School. In this study, 95% of the population 40 or more years old participated. Participants were first selected by postprandial glycosuria, and a 75g oral glucose tolerance test was carried out in glycosuric subjects (Shigeta *et al.*, 1983). The urban survey covered a random sample of subjects visiting a non-profit institution in Tokyo for a health check-up. Subjects were 4,000 males and 1,000 females, all of whom likewise had a 75g oral glucose tolerance test (Kitazawa *et al.*, 1983). The results were classified according to the criteria of the WHO (1980) definition and are shown in Table 21.1. A very high rate of impaired glucose tolerance was found in the urban study.

Epidemiology of diabetes among migrant Japanese

Second generation migrant Japanese in the United States were studied in King County, Washington, by Fujimoto and his group at the University of Washington. The second generation Japanese Americans (Nisei) whom they studied were in the age range 45-75 years, mean 62.0, in which NIDDM prevalence increases. In a random sampling of these subjects (229 males and 162 females) there was a very high rate of NIDDM (Table 21.2). Also, additional subjects with diabetes were found when given a 75g oral glucose tolerance test. They also found a very high rate of impaired glucose tolerance.

Second generation Japanese Brazilians (Brazil-Nisei) in the Sao Paulo area were surveyed through 'Centro de Estudos Nipo-Brasileiros' by Iunes and his group. They reported that 11.1% of males and 7.9% of females were already diagnosed. These results were very similar to those observed in Seattle (Table 21.2, line 1).

Thus, there is remarkable similarity between these two Nisei samples in the Americas. But there is also a large difference in the prevalence of diabetes between these second generation Japanese in the United States and Brazil on the one hand and Japanese in Japan on the other. The prevalence of impaired glucose tolerance is also increased in the US when compared to Japan.

Possible cause of this difference and future prevalence of diabetes in Japan

The number of patients with diabetes in Japan has been analysed from statistics of the Ministry of Health and Welfare (Health and

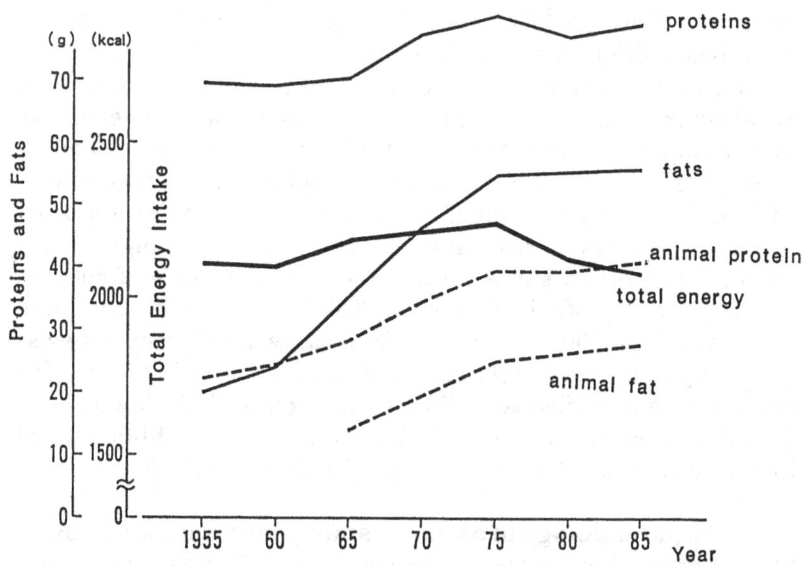

Figure 21.1. Changes in nutrition from 1955 to 1985.

Welfare Statistics Association, 1989). It appears that an increase occurred in the number of diabetic patients between 1955 and 1985 (Table 21.3). Annual rates of increase were calculated as mean yearly increase for each 5 year interval.

The number of patients with diabetes was 17,900 in 1955, and was about ten times that number fifteen years later. The high average rate of annual increase of diabetes (16%) over the first 15 years slowed between 1970 and 1975. This slow-down may be related to environmental problems that occurred then in Japan. During this period, the Japanese economy was seriously affected by the so-called 'first oil shock'. However, the high rate of increase returned in 1975 to 1980 when the mean annual rate of increase reached 19%, though it dropped in the succeeding quinquennium. If this rate were maintained into the early part of the next century, the prevalence of diabetes in the population older than 40 years would reach 20-25%. The cause of this rapid increase is not yet known. A factor of interest is nutrition. There was very little change in total energy consumed (Figure 21.1) *per caput*, over the 30 years 1955-85. However, there were dramatic increases in total fat, animal protein and animal fat intake that may be relevant.

There have been other environmental changes experienced in Japan associated with its westernisation. These include the

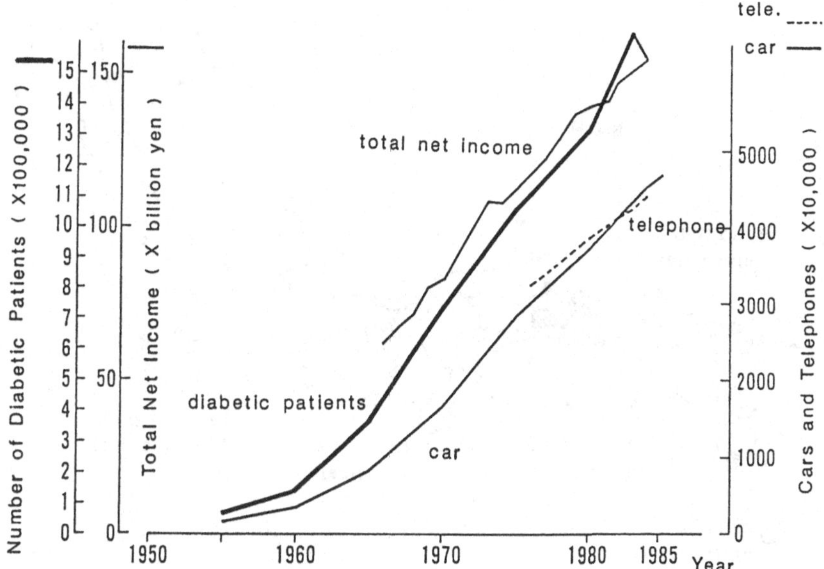

Figure 21.2. Changes in socioeconomic indices and number of diabetic patients in the period 1955-85.

increases in number of registered cars, number of telephones, and total net income in which many Nisei have participated. As illustrated in Figure 21.2, these three indices and number of diabetic patients appear to have increased almost in parallel. Although this association does not signify a causal relationship, it suggests that the disease incidence is related to some factor also associated with the measures in Figures 21.1 and 21.2.

Relatives in Japan of Seattle Nisei

Seattle Nisei were classified by the origin in Japan of their parents. There is some difference in prevalence of diabetes and impaired glucose tolerance by area of Japan from which their parents came (Leonetti & Fujimoto, 1989). This is illustrated in Figure 21.3.

This result prompted us to search for and examine members of their families who had remained in Japan. From the migration records the address and the village of origin of the parents could be identified. Using the koseki records in Japan it appeared possible to identify family members of persons who had been studied by Fujimoto's group. This koseki survey was carried out by Professor Norio Fujiki.

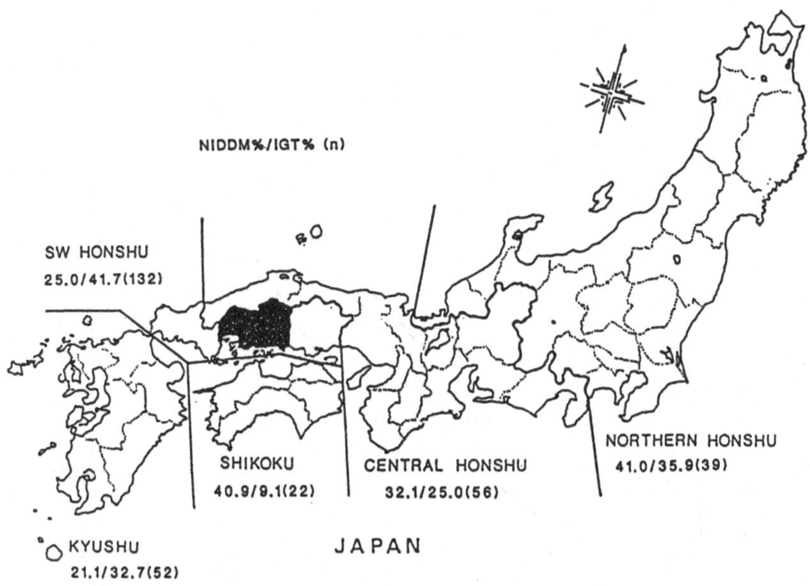

Figure 21.3. Prevalence of diabetes mellitus and impaired glucose tolerance in Seattle Nisei, classified by the origin in Japan of their parents.

Prior to the survey, it seemed that it would not be difficult to trace the persons required for the study. However, almost all of the addresses given in the koseki records were old. Many of the villages and towns in the records no longer exist or have changed their name so that addresses were difficult to find.

In spite of this difficulty, at least 10 persons who were cousins of Seattle Nisei have so far been identified. A letter requesting them to participate in the study was sent to each person. The letter consisted of a formal invitation from the head of the research group, an explanation in detail of our study including possible implications for the study of diabetes, and funding to cover the expenses of volunteers, including transportation. In addition to these, a letter from their cousin in Seattle was included. One letter of acceptance was received. Two letters were returned as having been sent to the wrong address. The other seven did not reply. The one person who responded was studied in a similar way to that in Seattle.

Thus, the approach of using a letter of invitation to participate was not sufficient to induce participation in this kind of study. The recipients may have been suspicious of a letter coming from an organisation with which they were unfamiliar, even if they were mailed from well-known university hospitals. To participate in this kind of study is uncommon in Japan, especially in rural areas. It is important to develop another method of approach to those who are potential participants. Probably, a more personal approach will be needed.

Conclusion

Environmental-genetic interaction appears to be important in the etiology of diabetes. A comparison of Japanese living in Japan with their relatives in other environments, Japanese born and reared in the United States and Brazil, is a potentially valuable approach to elucidate it. However, before such a comparison can be achieved, difficulties in identification and recruitment of relatives who are living in Japan must be resolved.

Acknowledgement

The authors are indebted to Professor Fumimaro Takaku, the 3rd Department of Internal Medicine, University of Tokyo, for his leadership in this work. This work was supported by a grant-in-aid from the Japan Society for Promotion of Science.

References

Barnett, A. H., Eff, E., Leslie, R. D. G. & Pyke, D. A. (1981). Diabetes mellitus in identical twins, A study of 200 pairs. *Diabetologia*, **20**, 87–93.

Fujimoto, W. Y., Leonetti, D. L., Kinyoun, J. L., Newell-Morris, L., Shuman, W. P., Storov, W. C. & Wahl, P. W. (1987a). Prevalence of diabetes mellitus and impaired glucose tolerance among second-generation Japanese-American men. *Diabetes*, **36**, 721–729.

Fujimoto, Y. Y., Leonetti, D. L., Kinyoun, J. L., Shuman, W. P., Stolov, W. C. & Wahl, P. W. (1987b). Prevalence of complications among second-generation Japanese-American men with diabetes, impaired glucose tolerance, or normal glucose tolerance. *Diabetes*, **36**, 730–739.

Health and Welfare Statistics Association: Trend of Health and Welfare of Japan (1989). *Kousei-no-Shihyou*, vol. **36**, Supplement.

Kitazawa, Y., Murakami, K., Goto, Y. & Hamazaki, S. (1983). Prevalence of diabetes mellitus detected by 75g GTT in Tokyo, Tohoku. *Journal of Experimental Medicine*, **141** (Suppl,) 229–234.

Kuzuya, T., Aoki, S., Isshiki, G., Okuyama, M., Kakizaki, M., Kadowaki, T., Jinnouchi, T., Hibi, I., Horino, M., Matsuda, A. & Miyamura, K. (1987). Diabetic Twins in Japan – Report of the committee on diabetic twins. *Journal of the Japanese Diabetes Society*, **30**, 1047–1063.

Leonetti, D. L. & Fujimoto, W. Y. (1989). Type 2 diabetes, impaired glucose tolerance, and hypertension in offspring of migrants and the structure of the population of origin. *Human Biology*, **61**, 369–387.

Shigeta, Y., Kikkawa, R., Kobayashi, N. & Katabami, J. (1983). A community study of diabetes in a population with a high diabetes mortality rate. *Tohoku Journal of Experimental Medicine*. **141** (Suppl.) 257–260.

W.H.O., Expert Committee on Diabetes Mellitus (1980). Second Report. *W.H.O. Technical Report Series*, No. 646.

22 *Environmental factors affecting ischemic heart disease*

GORO MIMURA, KEIJI MURAKAMI, MASAMICHI GUSHIKEN, SHOZO OGAWA

Most diseases, their occurrence and progression, are affected by genetic and environmental factors. In searching for genetic influences in disease it is important to investigate different races or groups living in similar environments. On the other hand, it is important also to simplify the many environmental factors when seeking environmental effects on the onset or progression of diseases in groups having similar genetic background. These two strategies of investigation, varying genetic constitution within similar environments and varying environments but constant genetic constitution, are complementary.

In ischemic heart disease (IHD), several factors such as hypertension (Kannel *et al.*, 1971a; WHO, 1982), hyperlipoproteinaemia (Keys *et al.*, 1958; Kinch *et al.*, 1963; Kannel *et al.*, 1971b), diabetes mellitus (Donahue *et al.*, 1987; Fuller *et al.*, 1980), obesity (Gordon & Kagan, 1981), and smoking habit (Mimura *et al.*, 1984) are thought to be related to its progression. These risk factors are influenced, at least in part, by environmental factors. Out of these environmental factors, dietary habit is suggested by many epidemiological studies to be one of the most important. Furthermore, dietary habit varies with the climate, geographical location, socioeconomic situation, and culture and history of the population.

This study compared the structure and background risk factors of IHD together with the nutritional situation in Okinawans and in migrants from Okinawa who were living in Honolulu without intermarriage (Okinawan-American). The leading causes of death are different, IHD and malignant neoplasia in Okinawan-American and Okinawan, respectively, in spite of their having the same genetic ancestry and living in climates of similar temperature (except in the winter season).

Subjects and experimental methods

Risk factors and IHD were surveyed in residents living in Honolulu city (USA) and Yonashiro town in Okinawa Prefecture (Japan). There were 1050 subjects (453 male, 597 female, average age: 56.3±14.4) in Honolulu and 1253 (538 male, 715 female, average age: 57.0±15.9) in Okinawa. These were measured and electrocardiograms obtained. Blood samples were obtained early in the morning after overnight fast, and the specimens were then kept frozen and sent to the University of the Ryukyus. All these samples were examined using the same methods. Serum total cholesterol, triglyceride, HDL-cholesterol, fasting plasma glucose, and uric acid were assayed and urinalysis carried out. Hypercholesterolaemia and hypertriglyceridaemia were defined when plasma cholesterol and triglyceride were more than 220mg/dl or 160mg/dl, respectively. IHD was diagnosed when the ECG pattern showed the I-1, IV-1,2, V-1,2 of the Minnesota code classification, and hypertension was diagnosed by the WHO criteria. The degree of obesity was expressed using the Body Mass Index, obesity being defined as a BMI of over 25. For the estimation of daily salt intake and diet, sodium output in a 24 hour urine specimen was measured, and a house to house nutritional survey by a dietician was done. Student's t-test and χ^2 were used in the statistical analysis and the results were expressed as mean±SD or percent values.

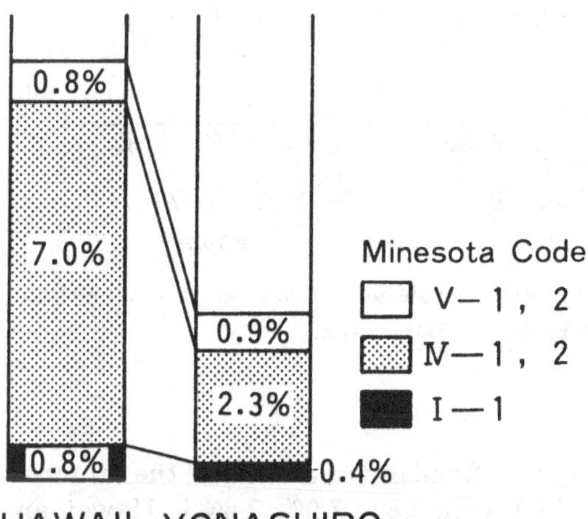

Figure 22.1. Prevalence of the ischaemic change of ECG in Okinawan-American and Okinawan.

Table 22.1. *Comparison of the prevalence of risk factors between Okinawan-American and Okinawan*

	Hawaii	*Yonashiro*	
Number	1050	1253	
Hypertension	11.9%	17.5%	p<0.01
Hypercholesterolaemia	54.0%	25.2%	p<0.01
Hypertriglyceridaemia	32.6%	22.8%	p<0.01
Obesity	37.5%	36.7%	N.S.
Smoking	15.6%	26.3%	p<0.01

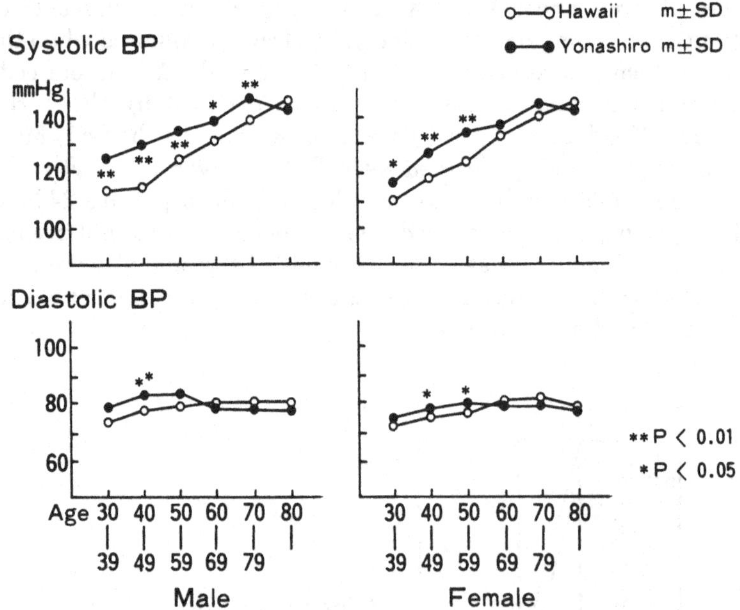

Figure 22.2. Systolic and diastolic blood pressure by sex and age in Okinawan-American and Okinawan.

Results

Prevalences of IHD as defined by the criteria of the Minnesota code of V-1,2, IV-1,2 and I-1 were 0.8%, 7.0%, 0.8% in Hawaii and 0.9%, 2.3%, 0.4% in Okinawa, respectively. As a whole ischemic change was found 2.5 times more frequently in the Hawaiian group (Figure 22.1). The prevalences of risk factors differ between the

Table 22.2. *Mean values of blood pressure, serum lipids and degree of obesity (BMI) in Okinawan-American and Okinawan*

		Hawaii	*Yonashiro*	
Systolic blood	M	127.0±17.1	136.0±21.6	p<0.01
presure	F	126.4±20.1	133.0±22.3	p<0.01
Diastolic blood	M	78.8±09.5	80.1±12.5	NS
pressure	F	77.5±09.5	78.2±12.9	NS
T. Cholesterol	M	226.4±43.8	193.0±34.6	p<0.01
	F	227.6±46.2	203.3±39.9	p<0.01
Triglyceride	M	189.0±08.3	124.9±03.7	p<0.01
	F	141.9±04.9	125.8±03.1	p<0.01
HDL-C	M	40.1±11.3	57.1±14.3	p<0.01
	F	46.5±13.2	57.5±12.1	p<0.01
Relative	M	24.8±03.0	23.4±03.6	p<0.01
weight	F	23.6±03.2	24.4±03.5	p<0.01

two populations (Table 22.1). There is significantly more hypercholesterolaemia and hypertriglyceridaemia in Okinawan-American but more hypertension and smoking in Okinawan, while there was no difference in the frequency of obesity. Mean values of the relevant measurements are shown in Table 22.2.

Systolic blood pressure was significantly higher at each decade except the 8th in Okinawan males, and from the 3rd to 5th decade in Okinawan females, than in the Okinawan-American (Figure 22.2). Serum total cholesterol, triglyceride and HDL-cholesterol values by sex and age are shown in Figures 22.3-22.5. In both sexes and all age decades, significantly increased serum cholesterol and triglyceride levels and decreased HDL-cholesterol levels were observed in Okinawan-American. Although the prevalences of obesity in the two populations were similar, differences between the mean values of the Body Mass Index (Table 22.2) and in the pattern of change by age in each sex (Figure 22.6) were observed, with higher BMI values occurring in older male Okinawan-American (over the 5th decade), and in younger female Okinawan (from the 3rd to 5th decade).

There was no striking difference in nutritional composition of food between the two groups except salt intake (Table 22.3). On

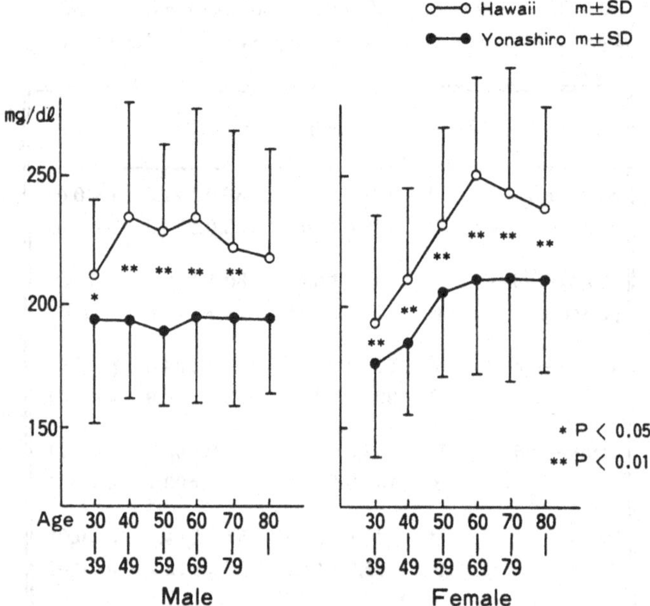

Figure 22.3. Serum cholesterol levels by sex and age in Okinawan-American and Okinawan.

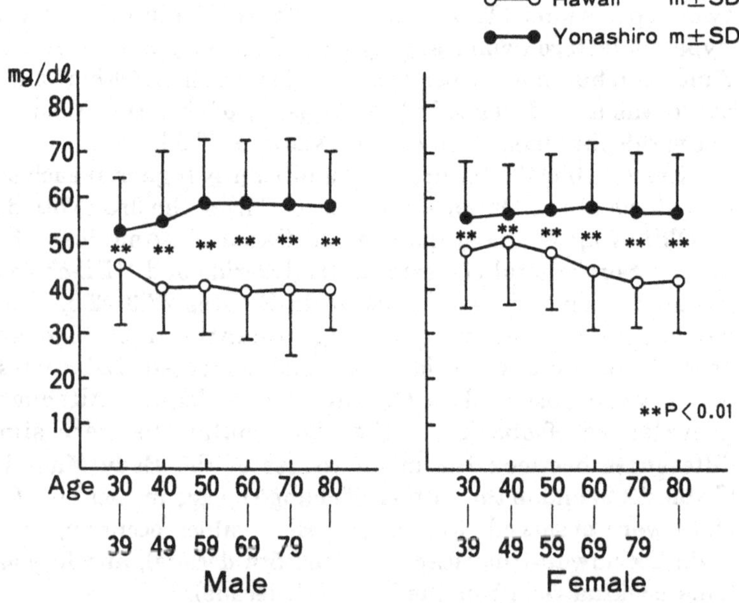

Figure 22.4. Serum HDL-cholesterol levels by sex and age in Okinawan-American and Okinawan.

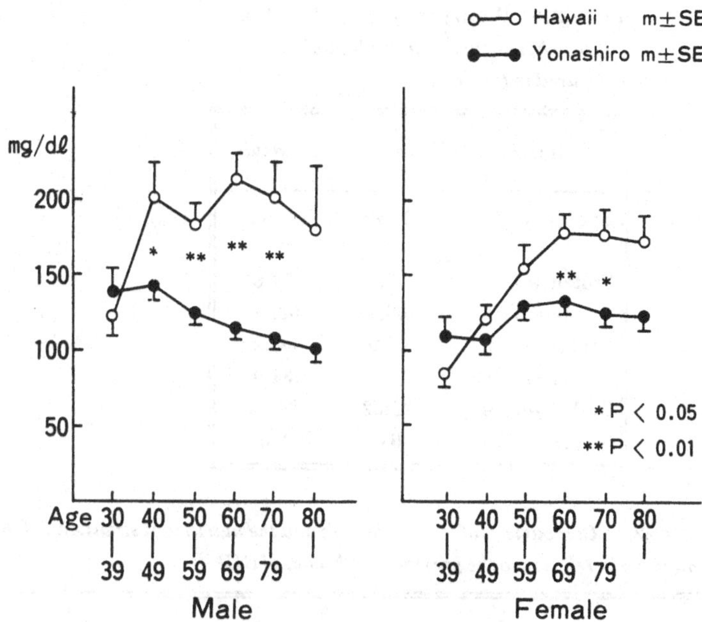

Figure 22.5. Serum Triglyceride levels by sex and age in Okinawan-American and Okinawan.

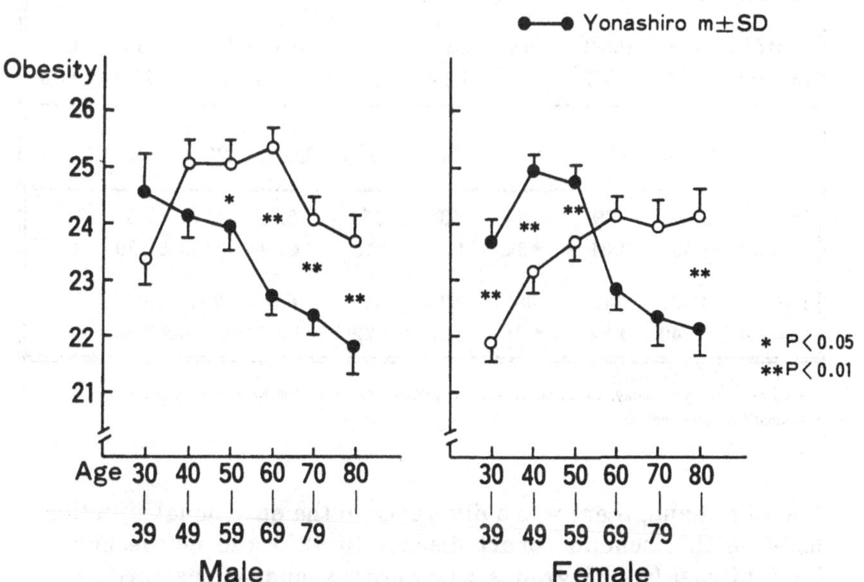

Figure 22.6. Relative body weight (BMI) of male and female in Okinawan-American and Okinawan.

Table 22.3. *Dietary composition based on house to house survey and salt intake measured in the urine*

Nutrient	*Hawaii*	*Yonashiro*
Energy (kcal)	1799	1866
Protein (g)	69.8	74.6
(animal)	(40.4)	(40.2)
Fat (g)	63.9	63.2
(animal)	(33.7)	(31.0)
Carbohydrate (g)	236.2	244.2
Na (g)	6.0	9.0

Table 22.4. *Comparison of the risk factors and the nutritional situation in Okinawan-American in relation to the presence or absence of IHD*

	Age	*Height*	*BW(kg)*	*BMI*	*HTN(%)*	*FPG*	*TC*	*TG*	*HDL-C*
IHD (-) (n=23)	66.8 ±7.0	155.7 ±7.1	56.8 ±11.4	23.2 ±3.2	13	94.9 ±6.4	210.3 ±32.1	136.6 ±73.0	53.4 ±9.8
IHD (+) (n=25)	66.3 ±9.8	155.2 ±9.5	60.9 ±10.1	25.2 ±2.4	60	106.2 ±41.7	207.6 ±35.8	155.7 ±71.5	46.1 ±11.4

	Energy	*CH*	*Pro.*	*Fat*	*Fat/En*	*Sodium*	*Chol.*	*P/S*
IHD (-) (n=23)	1575 ±495	209 ±31	67 ±34	53 ±27	37 ±18	5.6 ±1.3	170 ±139	1.23 ±0.30
IHD (+) (n=25)	1758 ±440	242 ±84	68 ±21	57 ±22	42 ±22	6.5 ±3.6	209 ±106	1.04 ±0.30

mean±SD, Energy:Kcal/day, CH:carbohydrate, Pro:protein, Fat:g/day, Fat/En:% of fat in total cal., Sodium:g/day, Chol:cholesterol

the other hand, there was a difference in the nutritional situation between the ischemic heart disease (n=25) and non-ischemic heart disease (n=23) groups, who were sex- and age-matched, with a lower P/S ratio and higher fat intake in the IHD group (Table 22.4).

Discussion

This study endeavoured to trace the role of environmental factors, especially of dietary habit, on the pathogenesis or the progression of IHD by comparing two populations who have the same genetic background, one of whom had changed its environment by migration from Okinawa to Hawaii. These two areas are similar in their geographical location in a subtropical zone and in socioeconomic standards. Therefore this study simplified the factors compared by restricting them to those that have changed as a result of the migration.

The results revealed a higher prevalence of IHD (8.6%) in Okinawan-American than in Okinawan (3.5%). In risk factors, the prevalence of hypercholesterolaemia and hypertriglyceridaemia was significantly higher in Okinawan-American, while there was a higher prevalence of hypertension in Okinawan who had a higher salt intake of 9.0g per day. The difference in the salt intake was presumed to be the reason for the difference in the blood pressure, for Houston (1986) reported a higher sensitivity of blood pressure to a moderate daily salt intake (sodium 50-100mEq/day) than to an excess intake. The salt intake in this study was very close to these values.

Keys *et al.* (1958) first reported the difference in the prevalence of both IHD and serum cholesterol levels between Japanese and Americans, interpreting the higher prevalence of IHD in the United States as deriving from dietary habits, rather than from any race difference. Since then there have been many studies, e.g. of Kagan *et al.* (1974) and dietary habits have changed as a result of information on the importance of reducing the intake of food containing cholesterol, animal fat or protein. The higher prevalence of hyperlipoproteinaemia in Okinawan-American and the lower P/S ratio observed in the IHD group in these populations may be related to these situations. Although hypertension has been regarded as one of the risk factors, the hyperlipoproteinaemia observed in Okinawan-American may be a more important background risk factor. The fact that Okinawan with their higher prevalence of hypertension and cigarette smoking showed a lower prevalence of IHD suggests a more important role of hyperlipoproteinaemia in the progression of IHD.

There is a recent tendency for serum cholesterol levels in Japanese to increase and in Americans to decrease. Therefore, through the long-term observation of these two populations, it may be possible to clarify more precisely the role of dietary habit in the progression of IHD.

References

Donahue, R. P., Abbott, R. D., Reed, D. M. & Yano, K. (1987). Postchallenge glucose concentration and coronary heart disease in men of Japanese ancestry. Honolulu Heart Program. *Diabetes*, **36**, 689–692.

Fuller, J. H., Shipley, M. J., Rose, G., Jarrett, R. J. & Keen, H. (1980). Coronary-heart-disease risk and impaired glucose tolerance. The Whitehall Study. *Lancet*, **i**, 1373–1376.

Gordon, T. & Kagan, A. (1981). Diet and relation to coronary heart disease and death in three populations. *Circulation*, **63**, 500–515.

Houston, M. C. (1986). Sodium and hypertension. A review. *Archives of Internal Medicine*, **146**, 179–185.

Kagan, A., Harrist, B. R., Winkelstein, W., Johnson, K. G., Kato, H., Syme, S. L., Rhoads, G. C., Gay, M. L., Nichaman, M. Z., Hamilton, H. B. & Tillotson, J. (1974). Epidemiologic studies of coronary heart disease and stroke in Japanese men living in Japan, Hawaii and California: demographic, physical, dietary and biochemical characteristics. *Journal of Chronic Disease*, **27**, 345–364.

Kannel, W. B., Castelli, W. P., Gordon T. & McNamara, P. M. (1971b). Serum cholesterol, lipoproteins, and risk of coronary heart disease. The Framingham study *Annals of Internal Medicine*, **74**, 1–12.

Kannel, W. B., Gordon, T. & Schwartz, M. J. (1971a). Systolic versus diastolic blood pressure and risk of coronary heart disease. The Framingham Study. *American Journal of Cardiology*, **27**, 335–346.

Keys, A., Kimura, N., Kusukawa, A., Bronte-Stewart, B., Larsen, N. & Keys M. H., (1958). Lessons from serum cholesterol studies in Japan, Hawaii and Los Angeles. *Annals of Internal Medicine*, **48**, 83–94.

Kinch, S. H., Doyle, J. T. & Hilleboe, H. E. (1963). Risk factors in ischemic heart disease. *American Journal of Public Health*, **53**, 438–442.

Mimura, G., Irei, M., Higa, S., Murakami, K., Ogawa, S. & Kagan, A. (1984). Epidemiological study on ischemic heart diseases in Okinawan and in Okinawan-American living in Honolulu. In *Nutritional prevention of cardiovascular disease*. pp. 241–250. Academic Press.

W.H.O. Expert Committee (1982). The prevention of coronary heart disease. Technical Report Series. pp. 678. W.H.O. Geneva.

Epilogue

DEREK F. ROBERTS

The foregoing pages concern the opposing processes of isolation and migration. In isolation a population maintains itself in (usually) a relatively constant environment, its gene pool remains unadulterated from outside, it evolves under the influences of natural selection, mutation, genetic drift and other random processes, and experiences internal changes in its genetic structure. Migration exposes the migrants to environmental change, so that comparison with those in the same population who did not migrate and are presumed to be of the same genetic constitution demonstrates the effect of change of environment, and so helps to separate out the genetic and environmental components of biological variation. This book aims to illustrate the biological effects of the two processes, to show how the situations to which they give rise may be used to elucidate a variety of biological problems ranging from evolutionary change to disease etiology. This object is achieved by a number of examples, the majority of which have been studied by Asian investigators and are less widely known than the classic Western studies of say the Amish, Yanomama or the Aland islanders (e.g. McKusick, 1978; Neel & Weiss, 1975; Eriksson, 1980). Besides documenting the results, the chapters illustrate the different methods employed in such studies.

Isolates

Isolation comes about or is maintained in many ways, and isolating factors include geographical barriers, distance, religious differences, hostility, mating patterns and other social and cultural variables (Roberts, 1975, 1984). The term isolate is used in many ways, ranging from the conceptual circle of marriages within an otherwise continuous population envisaged by Dahlberg (1929) to the small reproductively sequestered island populations where breeding and geographical boundaries coincide and are virtually absolute. Many workers in the past have been attracted to the study of small isolates, beginning long before the genetic significance of the isolating process was first formulated by Wahlund (1928). For they have many advantages for particular types of scientific study. The investigator is dealing with a finite and comprehensible population; indeed in the smaller isolates, and particularly those living on small islands, it is possible for the investigator to know every individual. The survey area also is often clearly defined within a circumscribed boundary. The population of an isolate moreover usually manifests a strong sense of identity, an identity usually supported by intricate internal genetic relationships.

There are several types of study to which isolates are particularly suited. A common feature of isolated populations is the occurrence of a few uncommon diseases of varying degrees of etiological complexity. These rarely exceed five or six in number and their combination is usually unique, each isolate showing a different group of diseases (e.g. Fujiki, this volume, Table 1.3). Isolate studies have brought to light numerous previously unknown disorders, which have helped us to understand different but related disorders elsewhere as with the X-linked Aland eye disease (Forsius & Eriksson, 1964) or Ryukyu spinal muscular atrophy (Kondo, this volume, p. ...). On account of the detailed information that can be obtained on the population, isolates therefore provide useful information on the etiologies. Sometimes these diseases reflect some ecological or cultural peculiarity, for example Kuru in the Fore of New Guinea (Gajdusek, 1977). More often these conditions are genetic, Mendelian or complex, and are the consequence of an historical accident, such as the arrival of an immigrant ancestor bearing a faulty gene, as with the Tristan syndrome (Roberts, 1980), or the occurrence of mutations in the population itself as with Osler Rendu disease in French Jura villages (Brunet, 1992).

The defined numbers of people in an isolated population and the finite area which they inhabit provide an excellent basis for prevalence estimates and subsequent epidemiological studies of particular disorders. Estimates made at different dates then bring to light secular trends in occurrence as with multiple sclerosis in Orkney (Poskanzer *et al.*, 1976)

The genetic constitution of isolated populations, as defined either by the frequencies of blood group and other alleles or by probable ancestral contributions to the gene pool, has proved of continuing interest. Small communities are particularly liable to random variations in gene frequency, proportionately greater than those which occur in a large population. Historical studies where the data exist allow the nature and extent of such random occurrences to be identified (e.g. Roberts, 1968). A major factor which makes genetic studies so appealing is that the pedigree of the isolated population is virtually closed and transmission of a gene can often be traced back over a number of generations, possibly to the initial founding of the population. It may thus be possible to trace absolutely, or by relative probabilities, which ancestors were responsible for introducing which genes (Thompson, 1978).

Migration

By contrast to isolation, the biological study of migration has received much less attention (Mascie-Taylor & Lasker, 1988; Roberts, 1988). It was only in the 1930s that it began to be realised that migration provided a kind of natural experiment by which the biological effect of change of environment could in theory be deduced. To do so in practice is not so simple, for it is necessary to establish that the emigrants were characteristic genetically of the population from which they derived and it became clear from quite early studies (e.g. Martin, 1949) that migrants differed from nonmigrants in a number of ways. However the principle remains and, provided that due attention is given to possible sources of error, migrant studies can be extremely revealing, as those incorporated in this volume show.

Migration, change in the location of residence, is quantitatively and qualitatively a highly heterogeneous process. It may concern a single individual who moves to seek employment or to marry; a group of individuals who are deported or recruited for labour or for military service. It may concern a family or group of families who seek a better life elsewhere or join their members who moved earlier as individuals. It may concern a subpopulation, particularly religious minority groups seeking religious freedom. Or it may be an intrinsic feature of the way of life of a total population, as with the gypsies and the nomadic and hunter-gatherer peoples. For the individuals concerned it may be temporary or permanent, recurrent, seasonal, or occur once in a lifetime. For the populations concerned it may occur in a single generation, or it may represent a continuing stream over several generations. It may be initiated by economic, social, political, psychological or religious pressures. It may be local (usually individuals or family) or it may be distant. Its biological effects and implications for the population of origin, for the existing inhabitants of the recipient area, and for the migrants themselves, depend on its scale and nature.

The studies of migration in this volume ignore the type of the migration, and are concerned only with those populations who have migrated or are in process of doing so, to show how their study can be used to understand etiologies. By contrast the chapters on isolation range rather more widely, and concern all types of isolates ranging from mating circles in a continuum to island geographical groups.

The book spans four topics, isolate dynamics, the biological features of isolates, diseases in isolates, and the use of migration studies. The background to their choice is shown by Fujiki in his introductory chapter, noting the particular advantages that Japan offered for isolate studies and her great contribution to them since the war, and how the clues to etiology of some diseases that these provided led to a further study of these diseases in those who had left their parental populations and moved elsewhere.

Isolate dynamics

In a thought-provoking chapter Neel differentiates the genetic situations covered by the term 'isolate' and argues that it is the way in which the population originated which determines the genetic effects. Yanase summarises the changes that are occurring in isolates in Japan, showing that not all are affected equally by the process of modernisation. Imaizumi enquires how far social factors explain the variations in consanguineous marriage frequency in Japan, and hence changes in isolating tendencies. Vogel considers some effects of the break-up of isolates that have received little attention. By contrast to the favourable outcome of reduction in frequency of homozygotes for deleterious recessives, a less favourable likely effect is disturbance of the precarious balance that there is between the genetic mechanisms of defence of the human host against infectious agents, and of attack of infecting organisms on the human host. This section explains why few overall generalisations can be made regarding the evolution of isolates. Since every isolate is in many senses unique, in its genetic constitution, its history, its size, its environment, its disease pattern, its exposure to external forces, the actual changes that are likely to occur will vary from one to another.

Biological features of isolates

The same theme but at a more descriptive level and concentrating on India is developed in the next three chapters. Malhotra examines the role of the social system in the origin and maintenance of isolates in India, using genetic evidence to establish that the castes represent the outcome of diversification from an original source rather than the agglomeration of unrelated elements. This theme is pursued by Mukherjee who concentrates on the consanguinity implicit in an endogamous caste structure and shows how it, and its genetic consequences, vary across India. Papiha relates higher levels of circulating immunogobulins in Indian tribal populations to selective stress and examines the diversification of the gene frequencies for the Ig allotypes that would be an expected result of differences in such stresses.

The arrival in Israel of Jewish immigrants, stemming from ancestral groups who settled in different parts of the world, provides an opportunity to examine the extent of genetic variation derived from their dispersion and relative isolation. Bonné-Tamir *et al.* compare the distances among them as assessed by DNA and serological genetic polymorphisms and finds similar patterns of affinity. The implication is a common genetic substratum for all Jewish peoples. Sukernik's analysis of Gm haplotypes in 31 circumpolar populations shows distinct groupings, with less genetic diversity in aboriginal populations in north-east Siberia, Asiatic Eskimos having the most improverished genetic array, and the Siberian Chukchi and American Athapaskan representing a continuum. The distinct Gm patterns observed are interpreted as the result of regional differentiation in north-east Asia, and comparison with data on American populations suggests that there were three distinct migrations from different Siberian homelands into the New World. These two chapters show an application of studies of polymorphisms in isolates additional to those noted by Fujiki (p. 11-12), helping to understand the history of dispersion and settlement of peoples.

The last two papers in this section concern advances in the methods of representing genetic data, and are ancillary not only to the other studies in this section but also to broader work in population genetics. Yasuda sets out algorithms for estimation of frequencies of haplotypes or tandemly linked loci. The depiction by Saitou *et al.* of the genetic relationship of populations by a network represents more appropriately than the usual dendrograms the evolutionary situation in human populations between whom migration occurs.

Diseases in isolates

A biological feature that is peculiar to many small isolated populations is the fact that various diseases, rare elsewhere, show higher incidences, so that such populations provide excellent opportunity for elucidation of etiologies. Kondo discusses the curious inherited and neurological disorders found in isolated populations in southern Japan, and shows the role of the feudal conditions in restricting their distribution. The study by Yanigihara and Garruto deals with infective rather than hereditary diseases. They show a focus of human T-lymphocyte virus type 1 (HTLV-1) infection in a remote isolate in Papua New Guinea and suggest that such foci are of ancient origin and are not attributable to recent introduction. A third type of disorder, the multifactorial, is the topic of investigation by Hamaguchi *et al.* of genes associated with hypercholesterolaemia by molecular analysis. Of the apolipoprotein genes, the same mutation of the ε4 allele occurs in Japanese and north-west Europeans, while the mutant genes for the low density lipoprotein receptor (LDL) that are responsible for familial hypercholesterolaemia show differing mutational origins from family to family. These chapters show the role of isolation, social and geographical, in restricting the spread of both genetic and infective disease, and raise the question of why mutations, shown by molecular studies to be the same, should occur in geographically diverse and distant populations.

Migration

The final group of chapters illustrates the complexity of migration and the difficulties that this poses in migrant studies; how simple data on migration help explain the pattern of normal biological variation; and the application of the studies of migrants to identify risk factors in a number of diseases Baker points out the usefulness of studies of migrants and the opportunity that they provide for understanding the underlying causes of variability among human populations, but cautions against simplistic interpretations. Prior's review of the study of migrants from Tokelau island to New Zealand summarises the development of the project and some of the results, and particularly the effect on body weight, diabetes, gout and blood pressure. In its broadest sense this is a study of the relationship between social change and health in this small Polynesian society. Lefevre-Witier's thoughtful paper considers the difficulties of defining groups, noting that the genetically defined population does not necessarily coincide with the ecologically defined group. He illustrates these discrepancies from his work on Tuareg. The balance between migration and isolation is explored by Rudan *et al.* among villages of the Dalmatian coastal area. The islands are shown to vary in their population structure, and simultaneous consideration of biological and cultural data traces the source of this variation to the differential effects of recent migrations.

Three Japanese studies attempt to identify risk factors in disease by comparing Japanese migrants to the USA who have adopted the American lifestyle with those who have remained in Japan. The comparison by Hara *et al.* of data on diabetes in Japanese migrants in Hawaii and Los Angeles and Japanese nonmigrants in Hiroshima Prefecture, shows a higher prevalence in Japanese Americans than in the sedentes, and in the former also a higher prevalence in second to fourth generation Japanese Americans (born in America) than in the first generation (immigrant Japanese from Japan). The differences are attributable to higher consumption of animal fat and less carbohydrate and less heavy physical activity, for laboratory studies of diabetics from both sources showed a significantly higher prevalence of hyperlipidaemia and higher mean LDL cholesterol level in the Japanese Americans. Kanazawa et al. compare the occurrence of non-insulin dependent diabetes in second generation Japanese migrants in Brazil and the USA with Japanese remaining in rural and urban Japan. There is a contrast between the migrants in the prevalence of the disorder and in the rate of impaired glucose tolerance. But the requisite genetic analysis has not yet started since despite the excellence of the koseki records it is proving difficult to trace in Japan the relatives of the migrants. Mimura *et al.* examine ischaemic heart disease. The results confirm the importance of dietary factors and suggest a critical role for hyperlipoproteinaemia.

Conclusion

The studies described in this book range over a number of years. Many of the points that they illlustrate are well known from other studies on other populations. But the drawing together of these examples, with investigations initiated from different points of view, allows broader appreciation of the utility of studies of isolated and migrant populations, and so perhaps will encourage others to pursue similar enquiries. Such studies are urgent, to be undertaken before isolates completely disappear and their characteristic features are swamped by the changes that are today occurring under the demands and stresses of living in the overcrowded modern world. The juxtaposition of these examples moreover allows a number of questions to be discerned that have not so far been asked or answered. Why do some mutations have a world-wide distribution and others not? How did the gene frequency variations seen in different populations come about?

What is the effect of migration not only on the immigrants themselves but on the peoples from whom they derive? There are many such questions, quite apart from all those others that are raised by serious consideration of each chapter.

References

Brunet, G. (1992). Hereditary elliptocytosis in the French Northern Alps. *Journal of Biosocial Science*, in press.

Dahlberg, G. (1929). Inbreeding in man. *Genetics*, **14**, 425.

Eriksson, A. W. (1980). Genetic studies on Aland. In *Population structure and genetic disorders*, ed. A. W. Eriksson, H. R. Nevanlinna, P. L. Workman & R. K. Norio, pp. 459–470. London: Academic Press.

Forsius, H. R. & Eriksson, A. W. (1964). Ein neues Augensyndrom mit X-chromsomaler Tranmission *Klin Mbl. Augenheilk.*, **144**, 447–457.

Gajdusek, D. C. (1977). *Unconventional viruses and the origin and disappearance of Kuru*, Stockholm: Nobel Foundation.

McKusick, V. A. (1978). *Medical genetic studies of the Amish*. Baltimore: Johns Hopkins University Press

Martin, W. J. (1949). *Physique of young adult males*. London: HMSO.

Mascie-Taylor, C. G. N. & Lasker, G. W. (1988). *Biological aspects of human migration*. Cambridge: Cambridge University Press.

Neel, J. V. & Weiss, K. M. (1975). The genetic structure of a tribal population, the Yanomama Indians. *American Journal of Physical Anthropology*, **42**, 25–52.

Roberts, D. F. (1968). Genetic effects of population size reduction. *Nature* (London), **220**, 1084–1088.

Roberts, D. F. (1975). Genetic studies of isolates. In *Modern trends in human genetics*, ed. A. E. H., Emery. pp. 221–269. Edinburgh: Butterworth.

Roberts, D. F. (1980). Genetic structure and the pathology of an isolated population. In *Population structure and genetic disorders*, ed. A. W. Eriksson & H. R., Forsius. pp. 7–26. London: Academic Press.

Roberts, D. F. (1984). Isolates and isolating factors. In *Human genetics and adaptation*, ed. K. C. Malhotra & A. Basu. pp. 347–362. Calcutta: Indian Statistical Institute.

Roberts, D. F. (1988). Migration and genetic change. *Human Biology*, **60**, 521–539.

Poskanzer, D. C., Walker, A. M., Yonkondy, J. & Sheridan, J. L. (1976). Studies in the epidemiology of multiple sclerosis in the Orkney and Shetland Islands. *Neurology*, **26**, 14–17.

Thompson, E. A. (1978). Inference of genealogical structure. *Social Science Information*, **15**, 477–526.

Wahlund, S. (1928). Zusammensetzung von Populationen und Korrelationserscheinungen vom Standpunkt der Vererbungslehre aus betrachtet. *Hereditas*, **11**, 65.

Index

Page numbers in bold refer to whole chapters.